M000272592

NUMBER POWER

A REAL WORLD APPROACH TO MATH

Transitions Math

McGraw Hill Education

Bothell, WA • Chicago, IL • Columbus, OH • New York, NY

www.mheonline.com

 **Education**

Copyright © 2011 by The McGraw-Hill Companies, Inc.

Cover Ellen Isaacs/Alamy.

Send all inquiries to:
Contemporary/McGraw-Hill
130 E. Randolph St., Suite 400
Chicago, IL 60601

ISBN: 978-0-07-661499-8
MHID: 0-07-661499-9

Printed in the United States of America.

2 3 4 5 6 7 8 9 10 RHR 15 14 13 12

TABLE OF CONTENTS

TO THE STUDENT

Number Power Transitions Math is designed to aid you in your transition to college-level math concepts. It assumes that you have successfully tackled the topics in the other Number Power books and are ready to move on. The book extends these topics and builds a connection between the concepts, using graphs and visuals to explain wherever possible.

Number Power Transitions Math will help prepare you for taking placement tests and for pursuing further education or training. It is intended to bridge any gaps between what you have already learned and what you need to know as you continue your education. The instruction is designed to help you move past memorizing rules to mastering key concepts as you pursue continued success in your studies.

Each chapter provides step-by-step instruction and extensive practice of skills and concepts. The icons shown at the right indicate math terminology, warning notes for commonly made mistakes, and thought-provoking questions and information. Each chapter ends with a chapter review to check your understanding. At the back of this book is an answer key for all the exercises and reviews. Checking the answer key after you have worked through a lesson will help you measure your progress. Also included in the back is a glossary of the mathematical terms used in this book.

Math Talk

Caution

Think About It

RULES AND PROPERTIES

Algebra is the language of operations. Everything you can do in algebra is based on the properties of numbers, operations, and equality. To be good at algebra, you need to understand and master the rules in this chapter. These rules and properties answer the question, "What am I allowed to do and when can I do it?"

Numbers and Operations

In this lesson, you will learn to

- Apply the commutative, associative, and distributive properties.
- Use factoring and canceling to simplify expressions.
- Apply the order of operations to write and evaluate expressions.
- Solve problems involving the arithmetic average (mean).

Rules and properties are written with variables. Using variables in a rule shows that the rule will work for different values. When you read a rule, try substituting numbers for the variables to see how the rule works.

The **commutative property** states that you can add or multiply in any order.

RULE	EXAMPLES		
$a + b = b + a$	$3 + 4 = 4 + 3$	$9 + (-5) = (-5) + 9$	$x + (-y) = (-y) + x$
$ab = ba$	$5 \times 6 = 6 \times 5$	$-1(2) = 2(-1)$	$x \cdot 3y = 3xy$

The **associative property** states that you can group numbers in any way when you are adding or when you are multiplying.

RULE	EXAMPLES	
$a + (b + c) = (a + b) + c$	$(14 + 12) + 8 = 14 + (12 + 8)$ $26 + 8 = 14 + 20$ $34 = 34$	Changing the grouping can make it easier to add or multiply numbers mentally.
$a(bc) = (ab)c$	$(18 \cdot 5) \cdot 2 = 18 \cdot (5 \cdot 2)$ $90 \cdot 2 = 18 \cdot 10$ $180 = 180$	Look for ways to make factors like 10 or 100 that are easy to multiply by.

Combine the commutative and associative properties to make it easier to evaluate expressions.

EXAMPLE 1 Evaluate: $\left(\frac{1}{3} \times 8\right) \times 3$

STEP 1 Change the order.

$$\left(\frac{1}{3} \times 8\right) \times 3 = \left(8 \times \frac{1}{3}\right) \times 3$$

STEP 2 Change the grouping.

$$\left(8 \times \frac{1}{3}\right) \times 3 = 8 \times \left(\frac{1}{3} \times 3\right)$$

STEP 3 Multiply.

$$8 \times \left(\frac{1}{3} \times 3\right) = 8 \times 1 = 8$$

The distributive property can be illustrated by finding the area of a rectangle.

EXAMPLE 2 In the diagram at the right, two adjacent rectangles have the same width. Find the area of the large rectangle formed by the two smaller rectangles.

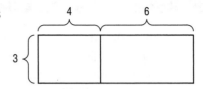

Remember To find the area of a rectangle, multiply length and width: $A = lw$.

There are two ways you can use the information to find the area of the rectangle.

Method 1 Find the area of each smaller rectangle by multiplying and then add.

4		6
3	12	18	3

$3(4) + 3(6) = 12 + 18$
$= 30$ square units

Method 2 Add the lengths of the smaller rectangles and then multiply by the width.

4 + 6 = 10
3	30

$3(4 + 6) = 3(10)$
$= 30$ square units

This leads to the **distributive property.**

RULE	EXAMPLES	
$a(b + c) = ab + ac$	$3(7 + 6) = 3(7) + 3(6)$	$6(x - 4) = 6x - 6(4)$ $= 6x - 24$

Factoring applies the distributive property in reverse. When the terms in an addition or a subtraction expression have a common factor, you can apply the distributive property to factor out that number.

EXAMPLE 3 Factor: $18y + 24$

The terms are $18y$ and 24. Both terms have the common factor 6. In other words, both terms can be divided by 6. Factor out the 6.

$18y + 24 = 6(3y) + 6(4)$
$= 6(3y + 4)$

Canceling is a shortcut for simplifying fractions. It works because any number (or variable) divided by itself equals 1.

To cancel, cross out the same number in the numerator and denominator of a fraction or a product of fractions.

$\frac{5}{8} \times \frac{8}{11} = \frac{5}{\cancel{8}} \times \frac{\cancel{8}}{11} = \frac{5}{11}$

 Caution: You cannot cancel when fractions or terms are added or subtracted.

$\frac{2}{\cancel{3}} + \frac{\cancel{3}}{5}$ does NOT equal $\frac{2}{5}$.

Now apply canceling to an algebraic expression. Remember, you are dividing both the numerator and denominator by the same number or term, so you are really dividing by 1.

EXAMPLE 4   Simplify: $\dfrac{18xy}{2y}$

STEP 1 Factor out $2y$ from the numerator and the denominator of the fraction.

$$\frac{18xy}{2y} = \frac{2y(9x)}{2y}$$

You can also cancel this way:

$$\frac{\overset{9}{\cancel{18}}x\cancel{y}}{\underset{1}{\cancel{2}}\cancel{y}} = 9x$$

STEP 2 Cancel $2y$ from the numerator and denominator. Remember, $\dfrac{2y}{2y}$ equals 1, so the value of the expression has not changed.

$$\frac{\cancel{2y}(9x)}{\cancel{2y}} = 9x$$

Use the listed property to write an equivalent expression.

1. Commutative $3 + 6$ $9(-2)$ $(-7) + 9$

2. Associative $(3 \cdot 5) \cdot 6$ $-2 + (6 + 10)$ $5(-4 \cdot -9)$

3. Distributive $9(8 - 12)$ $-2(x + 5)$ $3a(9 - b)$

Factor each expression using the distributive property.

4. $12a + 8b$ $21x - 7$ $6x + 12y$

5. $10a - 5b + 15$ $-3x - 6y$ $10x + 2xy$

Simplify by canceling.

6. $\dfrac{9xy}{3y}$ $\dfrac{18ab}{6ab}$ $\dfrac{3}{10} \cdot \dfrac{5}{18}$

7. $\dfrac{14}{x} \cdot \dfrac{x}{7}$ $\dfrac{2}{3x} \cdot \dfrac{9}{6y} \cdot \dfrac{xy}{8}$ $\dfrac{4a}{c} \cdot \dfrac{12}{25} \cdot \dfrac{5bc}{15a}$

The value of an expression can be affected by the order in which the operations are performed. To avoid confusion, mathematicians have developed an **order of operations** to use when finding the value of any expression.

The Order of Operations

1. Do any operations in parentheses first.
2. Evaluate all exponents and roots.
3. Do any multiplication and division, working from left to right.
4. Do any addition and subtraction, working from left to right.

EXAMPLE 5   Evaluate: $6 + 10 \cdot 4$

STEP 1 There are no parentheses or exponents, so do the multiplication first.

$$6 + 10 \cdot 4 = 6 + \underline{10 \cdot 4}$$
$$= 6 + 40$$

STEP 2 Then add.

$$= 46$$

EXAMPLE 6   Evaluate: $4(9 - 6)^2 - 12$

STEP 1 Subtract inside the parentheses first: $9 - 6 = 3$.

$$4(9 - 6)^2 - 12 = 4(3)^2 - 12$$

STEP 2 Then evaluate the exponent: $3^2 = 3 \cdot 3 = 9$.

$$= 4(9) - 12$$

STEP 3 Multiply.

$$= 36 - 12$$

STEP 4 Subtract.

$$= 24$$

Find the value of the expressions using the order of operations.

8. $3 \cdot 4^2 - 7 \cdot 3$ $36 \div (8 - 4) + 2$ $3(2 + 3)^2 - 8$

9. $7 + (3 \cdot 2 + 1)^2$ $\dfrac{(1 + 2 \cdot 4)^2}{3}$ $\left[\dfrac{7 + 2(5 - 4)}{3}\right]^2 \leftarrow$

> The brackets act like a second set of parentheses. Work from the inside to the outside.

Write *True* if a statement is true as written. If the statement is false, insert parentheses to make it true.

10. $6 \cdot 4 - 2 + 3 = 15$ $15 - 3 \cdot 4 = 3$ $7 - 4^2 - 1 = 8$

11. $4 + 2 \cdot 2^2 = 64$ $3^2 \cdot 8 - 7 + 2^2 = 81$ $16 \div 4 - 2 \cdot 3 \div 6 = 3$

Synthesis   Finding the **arithmetic average** (also called the **mean**) of a set of values involves both addition and division. You add the values in the set and then divide your answer by the number of values in the set. Using the order of operations, you can write the arithmetic average as a single expression.

Write an expression for finding the average of each group of numbers. Then find the value of the expression. The first problem has been started for you.

12. Test scores: 82, 96, 90, and 84   Prices: \$7, \$7, \$4, \$7, \$15

$$\dfrac{82 + 96 + 90 + 84}{4} =$$

13. Points scored: 16, 25, 19, 21, 31, and 14   Miles driven: 140, 390, and 220

Working with Exponents

In this lesson, you will learn to

- Find the value of expressions containing exponents.
- Apply the rules of exponents to simplify expressions.
- Use exponents to solve problems with scientific notation.

Exponents are used to express the repeated multiplication of a number. For example, $2^5 = 2 \cdot 2 \cdot 2 \cdot 2 \cdot 2$. In this example, 2 is the **base** and 5 is the **exponent.** You read the expression 2^5 as "2 to the 5th **power.**" The exponent tells how many times to use the base as a factor.

To evaluate an expression containing an exponent, perform the multiplication.

EXAMPLE 1 Evaluate: $3^4 = \underbrace{3 \cdot 3 \cdot 3 \cdot 3}_{4 \text{ times}} = \textbf{81}$

Memorize the following two special situations:

RULE	IN WORDS	EXAMPLE
$a^1 = a$	When any base is raised to a power of 1, the expression equals the base.	$10^1 = 10$
$b^0 = 1$	When any base is raised to a power of 0, the expression equals 1.	$10^0 = 1$

What happens when you raise a negative integer to a power?

EXAMPLE 2 Evaluate: $-6^2 = -(6 \times 6) = \textbf{−36}$

Evaluate: $(-6)^2 = (-6) \times (-6) = \textbf{36}$

Take another look. Make sure you understand how these are different.

- The expression -6^2 means *the negative of 6 squared*.
- The expression $(-6)^2$ means *the square of negative 6*.

> **Math Talk**
> A number raised to the 2nd power is said to be **squared.** A **cubed** number is raised to the 3rd power.

Now you're ready to review the laws of exponents. Remember, variables are used to show that the rules apply to any real numbers that can be put in place of the variables. Study each rule to find the idea it expresses.

RULE	IN WORDS	EXAMPLE
$a^b \times a^c = a^{b+c}$	When like bases are multiplied, keep the base the same and add the exponents.	$2^2 \times 2^3 = 2^{2+3} = 2^5$

Why does it work? $2^2 = 2 \cdot 2 \qquad 2^3 = 2 \cdot 2 \cdot 2$

Therefore: $2^2 \times 2^3 = (2 \cdot 2) \times (2 \cdot 2 \cdot 2) = 2^5$

RULE	IN WORDS	EXAMPLE
$(a^b)^c = a^{bc}$	When a base with an exponent is raised to a power, keep the base the same and multiply the exponents.	$(x^2)^3 = x^{2 \times 3} = x^6$

Why does it work? $(x^2)^3 = x^2 \cdot x^2 \cdot x^2 = x^{2+2+2} = x^6$
The base x^2 is multiplied three times.

Evaluate. Find the value of each expression.

1. 2^5 $\qquad\qquad\qquad$ 4^3 $\qquad\qquad\qquad$ -3^2

2. $(-4)^0$ $\qquad\qquad\qquad$ $(-2)^6$ $\qquad\qquad\qquad$ 14^1

3. $-2^3 + 3^2$ $\qquad\qquad\qquad$ $8^2 - 4^0$ $\qquad\qquad\qquad$ $(-5)^2 + (-5)^0$

Simplify. Express using exponents.

4. $3^2 \cdot 3^5$ $\qquad\qquad\qquad$ $x^3 \cdot x^5$ $\qquad\qquad\qquad$ $(-y)^1 \cdot (-y)^3$

5. $(a^2)^5$ $\qquad\qquad\qquad$ $2^8 \cdot 2^2 \cdot 2^3$ $\qquad\qquad\qquad$ $(-b^0)^3$

6. $a^3 \cdot a^4 \cdot a^1$ $\qquad\qquad\qquad$ $(4^2)^6$ $\qquad\qquad\qquad$ $(12^3)^5$

The next rule can be illustrated by canceling.

EXAMPLE 3 Simplify: $\dfrac{x^6}{x^4} = \dfrac{\cancel{x} \cdot \cancel{x} \cdot \cancel{x} \cdot \cancel{x} \cdot x \cdot x}{\cancel{x} \cdot \cancel{x} \cdot \cancel{x} \cdot \cancel{x}} = \dfrac{x^2}{1} = \mathbf{x^2}$

Instead of writing out the canceling step, you could subtract to find the answer.

RULE	IN WORDS	EXAMPLE
$\dfrac{a^b}{a^c} = a^{b-c}$	When like bases are divided, keep the base the same and subtract the exponents.	$\dfrac{4^5}{4^3} = 4^{5-3} = 4^2$

Now put the greater power in the denominator. What happens when you subtract?

EXAMPLE 4 Simplify by canceling. $\dfrac{x^4}{x^6} = \dfrac{\cancel{x} \cdot \cancel{x} \cdot \cancel{x} \cdot \cancel{x}}{\cancel{x} \cdot \cancel{x} \cdot \cancel{x} \cdot \cancel{x} \cdot x \cdot x} = \dfrac{1}{x^2}$

Simplify by subtracting exponents. $\dfrac{x^4}{x^6} = x^{4-6} = \mathbf{x^{-2}}$

Therefore, $\dfrac{1}{x^2}$ and x^{-2} are equal. This leads to another rule.

RULE	IN WORDS	EXAMPLE
$a^{-b} = \dfrac{1}{a^b}$	Moving a base with its exponent to the other side of the fraction bar changes the sign of the exponent.	$2^{-3} = \dfrac{1}{2^3}$

EXAMPLE 5 Express using positive exponents: $\dfrac{a^2 b^{-3} c^4}{a^3 b^2 d^{-1}}$

STEP 1 Subtract exponents with like bases.

$$\dfrac{a^{(2-3)} b^{(-3-2)} c^4}{d^{-1}} = \dfrac{a^{-1} b^{-5} c^4}{d^{-1}}$$

STEP 2 Move variables with negative exponents to the other side of the fraction bar and change the exponents to positive.

$$\dfrac{a^{-1} b^{-5} c^4}{d^{-1}} = \dfrac{c^4 d}{ab^5}$$

The final rule of exponents is used in multiplication and division problems. You can think of it as distributing the exponent.

RULE	IN WORDS	EXAMPLE
$(ab)^c = a^c \times b^c$ $\left(\dfrac{a}{b}\right)^c = \left(\dfrac{a^c}{b^c}\right)$	When a product or a quotient is raised to a power, the exponent is applied to each item in the product or quotient.	$(2y)^2 = 2^2 \times y^2 = 4y^2$ $\left(\dfrac{2}{3}\right)^2 = \left(\dfrac{2^2}{3^2}\right) = \dfrac{4}{9}$

 Caution: You cannot distribute an exponent when adding or subtracting.

$(x + y)^2$ does NOT equal $x^2 + y^2$.

Evaluate. Find the value of each expression.

7. $\dfrac{3^5}{3^3}$ $\qquad\qquad$ $\left(\dfrac{3}{4}\right)^2$ $\qquad\qquad$ $(-3)^{-2}$

8. 5^{-3} $\qquad\qquad$ $\dfrac{7^5}{7^4}$ $\qquad\qquad$ $\left(\dfrac{4^2}{4^4}\right)^{-1}$

Simplify. Express using positive exponents.

9. $(5a)^2$ $\qquad\qquad$ $(3x^2)^3$ $\qquad\qquad$ $(ab^2 c^3)^2$

10. $(y^2)^5$ $\qquad\qquad$ $(xy^{-2})^3$ $\qquad\qquad$ $(-4)^{-2} b^3$

11. $a^0 b^{-3} c^2$ $\qquad\qquad$ $\dfrac{6x^4}{3x^3}$ $\qquad\qquad$ $(x^3 y^2 z^{-1})^{-2}$

Write each expression (a) using negative exponents and (b) using positive exponents.

12. $\dfrac{a^3}{a^7}$ $\qquad\qquad$ $\dfrac{y}{y^4}$ $\qquad\qquad$ $\dfrac{m^{-3}n^2}{m^2n^4}$

Write each expression using only positive exponents.

13. $\dfrac{m^{-3}}{n^{-2}}$ $\qquad\qquad$ $\dfrac{x^2y^{-3}}{x^5}$ $\qquad\qquad$ $\dfrac{2^3a^{-2}}{2^{-2}a^5b^{-3}}$

Synthesis In **scientific notation,** very small or very large values are written as the product of two factors. The first factor is 1 or a number between 1 and 10. The second factor is a power of 10 written in exponential form. Watch for the relationship between the power of 10 and the placement of the decimal point.

EXAMPLE 6 $\quad 5.4 \times 10^5 = 540{,}000 \qquad$ The decimal point moved 5 places to the right.

$1.2 \times 10^{-4} = 0.00012 \quad$ The decimal point moved 4 places to the left.

You can use the rules of exponents to multiply or divide numbers written in scientific notation.

Simplify. Write your answer in scientific notation. The first steps are done for you.

14. $(2.5 \times 10^2) \times (2.0 \times 10^3)$
$= 2.5 \times 2.0 \times 10^2 \times 10^3$
$= (2.5 \times 2.0) \times (10^2 \times 10^3)$
$=$

15. $\dfrac{9.0 \times 10^6}{3.0 \times 10^2} = \dfrac{\overset{3}{\cancel{9.0}} \times 10^6}{\underset{1}{\cancel{3.0}} \times 10^2} =$

Simplify the expressions. Write your answers in scientific notation.

16. $(1.5 \times 10^3)(3.0 \times 10^6)$ $\qquad\qquad$ $(2.0 \times 10^{-5})(3.4 \times 10^2)$

17. $\dfrac{8.4 \times 10^8}{2.0 \times 10^3}$ $\qquad\qquad$ $\dfrac{3.0 \times 10^4}{1.5 \times 10^{-2}}$

18. Challenge Light travels 1.86×10^5 miles per second. It takes light approximately 5.0×10^2 seconds to reach Earth from the sun. What is the distance in miles from Earth to the sun expressed in scientific notation?

Working with Radicals

In this lesson, you will learn to

- Evaluate expressions containing radicals.

- Apply the rules of radicals when working with expressions.

- Express radicals in simplest form.

Radicals are closely related to exponents. You know $5^2 = 25$, read *5 squared equals 25.* You also know that $\sqrt{25} = 5$, which is read: *the square root of 25 is 5.* Study the two expressions and think about how they are related. The radical symbol reverses the exponent process.

EXAMPLE 1 What is the value of $\sqrt{16}$?

To solve the problem, you need to ask yourself, "What number can I square to get 16?" You know that $4^2 = 16$ and $(-4)^2 = 16$. Therefore, 16 has two square roots: **4** and **−4.** In fact, every positive number has two square roots.

To avoid confusion, the positive square root is called the principal root: $\sqrt{16} = 4$. Write a negative sign to indicate the negative square root of a number: $-\sqrt{16} = -4$.

Radicals are also used when the exponent is greater than 2. You already know that $2^3 = 8$. Now reverse the process: $\sqrt[3]{8} = 2$, which is read: *the cube root of 8 is 2.* The exponent 3 becomes a small index number written next to the radical sign.

EXAMPLE 2

$\sqrt[4]{1} = ?$
What is the fourth root of 1?
$1 \cdot 1 \cdot 1 \cdot 1 = 1$, so $\sqrt[4]{1} = \mathbf{1}$.

$\sqrt[3]{-27} = ?$
What is the cube root of −27?
$(-3)(-3)(-3) = -27$, so $\sqrt[3]{-27} = \mathbf{-3}$.

Were you surprised to see a negative number in the radical symbol? You can find a root of a negative number as long as the index is an odd number.

> ⚠ **Caution:** An even root of a negative number is not a **real number.** $\sqrt{-8} = ?$
> What real number multiplied by itself equals −8? There isn't one. The solution to this expression is an **imaginary number.** The concept of imaginary numbers is beyond the scope of this book.

Find the value of each expression.

1. $\sqrt{49}$ $-\sqrt{144}$ $\sqrt{36}$ $-\sqrt{81}$

2. $\sqrt[4]{16}$ $\sqrt[3]{125}$ $\sqrt[3]{-64}$ $\sqrt[5]{-1}$

As you consider the properties of radicals, think about how they relate to the rules of exponents. The first rule is used to simplify expressions with radicals. It helps to separate the root into factors so that you can simplify part of the root.

RULE	IN WORDS	EXAMPLE
$\sqrt{ab} = \sqrt{a} \cdot \sqrt{b}$	The root of a number is equal to the product of the roots of its factors.	$\sqrt{18} = \sqrt{9} \cdot \sqrt{2}$

EXAMPLE 3 Simplify: $\sqrt{45}$

STEP 1 Look for a factor of 45 that is a perfect square.

STEP 2 Find the root of the perfect square.

$$\sqrt{45} = \sqrt{9} \cdot \sqrt{5}$$
$$= 3\sqrt{5}$$

> Perfect squares are formed when you square a whole number.
>
> $3^2 = 9$, so 9 is a perfect square.

In the following example, you will simplify an expression where a number is multiplied by the root.

EXAMPLE 4 Simplify: $3\sqrt{32}$

STEP 1 Look for a factor of 32 that is a perfect square.

STEP 2 Find the root of the perfect square.

STEP 3 Multiply.

$$3\sqrt{32} = 3 \cdot \sqrt{16} \cdot \sqrt{2}$$
$$= 3 \cdot 4 \cdot \sqrt{2}$$
$$= 12\sqrt{2}$$

The second rule of radicals shows how to handle division and radicals.

RULE	IN WORDS	EXAMPLE
$\sqrt{\dfrac{a}{b}} = \dfrac{\sqrt{a}}{\sqrt{b}}$	To find the root of a fraction, find the roots of the numerator and denominator separately.	$\sqrt{\dfrac{36}{49}} = \dfrac{\sqrt{36}}{\sqrt{49}} = \dfrac{6}{7}$

EXAMPLE 5 Simplify: $\sqrt{\dfrac{18}{81}}$

STEP 1 Find the roots of the numerator and denominator separately.

$$\sqrt{\frac{18}{81}} = \frac{\sqrt{18}}{\sqrt{81}} = \frac{\sqrt{9} \cdot \sqrt{2}}{9} = \frac{3\sqrt{2}}{9}$$

STEP 2 You can divide the numerator and denominator by 3. Canceling is shown here.

$$= \frac{\overset{1}{\cancel{3}}\sqrt{2}}{\underset{3}{\cancel{9}}} = \frac{\sqrt{2}}{3}$$

Simplify each expression. To simplify cube roots, look for factors that are perfect cubes.

3. $\sqrt{50}$ $\sqrt{54}$ $\sqrt{300}$

4. $\sqrt[3]{108}$ $\sqrt[3]{24}$ $\sqrt[3]{54}$

5. $4\sqrt{27}$ $3\sqrt{80}$ $2\sqrt{75}$

6. $3\sqrt{192}$ $2\sqrt[3]{40}$ $2\sqrt[3]{448}$

7. $\sqrt{\dfrac{1}{16}}$ $\sqrt{\dfrac{4}{25}}$ $\sqrt{\dfrac{49}{121}}$

8. $\sqrt{\dfrac{3}{100}}$ $\sqrt{\dfrac{18}{32}}$ $\sqrt{\dfrac{125}{25}}$

An expression is not in simplest form if there is a radical in the denominator of a fraction. You can use this basic idea to eliminate the radical: $\sqrt{a} \cdot \sqrt{a} = a$. This process is also known as **rationalizing the denominator.**

EXAMPLE 6   Simplify: $\sqrt{\dfrac{2}{5}}$

STEP 1   Work with the numerator and denominator separately. $\sqrt{\dfrac{2}{5}} = \dfrac{\sqrt{2}}{\sqrt{5}}$

STEP 2   The goal is to eliminate the radical in the denominator, so multiply by $\dfrac{\sqrt{5}}{\sqrt{5}}$. $\dfrac{\sqrt{2}}{\sqrt{5}} \cdot \dfrac{\sqrt{5}}{\sqrt{5}} = \dfrac{\sqrt{10}}{5}$

The expression $\dfrac{\sqrt{10}}{5}$ is in simplest form.

Why does the process of rationalizing the denominator work? When you multiply both the numerator and the denominator of a fraction by the same value, you are actually multiplying by 1. The fraction looks different, but its value is the same.

Simplify each expression.

9. $\sqrt{\dfrac{2}{3}}$ $\sqrt{\dfrac{7}{12}}$ $\sqrt{\dfrac{3}{8}}$

10. $\sqrt{\dfrac{5}{18}}$ $\sqrt{\dfrac{3}{50}}$ $\sqrt{\dfrac{7}{24}}$

The properties of radicals also work if the expression contains variables.

EXAMPLE 7   Simplify: $\sqrt{18x^3}$

STEP 1   Factor by looking for perfect squares. Both 9 and x^2 are perfect squares. The square root of 9 is 3, and the square root of x^2 is x.

$$\sqrt{18x^3} = \sqrt{9 \cdot x^2 \cdot 2x}$$
$$= 3x\sqrt{2x}$$

STEP 2   Write the square roots of the perfect squares as a product outside the radical symbol.

Simplify each expression.

11. $\sqrt{x^2y^2}$ $\sqrt{72a^5}$ $\sqrt{16m^4n^2}$

12. $\sqrt{\dfrac{12n^2}{25}}$ $\sqrt{\dfrac{2a^2}{7}}$ $\sqrt{\dfrac{m^5}{3n^2}}$

Synthesis   Consider these two examples: $3^{\frac{1}{2}} = \sqrt{3}$ and $x^{\frac{2}{3}} = \sqrt[3]{x^2}$. What relationship do you notice between the **fractional exponent** and the radical expression? The denominator of the fraction is the index number of the radical. The numerator is the exponent of the number inside the radical. Using this relationship, you can perform operations with radicals using the rules of exponents.

EXAMPLE 8   Multiply: $\sqrt{2} \cdot \sqrt[4]{2}$

STEP 1   Write the radicals using exponents.

$$2^{\frac{1}{2}} \cdot 2^{\frac{1}{4}}$$

STEP 2   The bases are the same, so keep the base and add the exponents. Notice that you have to rewrite the fractions with a common denominator before adding.

$$2^{\frac{1}{2}+\frac{1}{4}} = 2^{\frac{2}{4}+\frac{1}{4}} = 2^{\frac{3}{4}}$$

STEP 3   Write the result as a radical.

$$2^{\frac{3}{4}} = \sqrt[4]{2^3}$$

Simplify by rewriting with fractional exponents.

13. $\sqrt{5} \cdot \sqrt[3]{5}$ $\sqrt[3]{x} \cdot \sqrt[4]{x}$

14. $\left(\sqrt[3]{4^2}\right)^3$ $\left(\sqrt{m}\right)^3$

Rules and Properties Review

Identify the property used in each expression.

1. $-4 + (x - 2) = (-4 + x) - 2$
   a. commutative
   b. associative
   c. distributive

   $7(3 - y) = 21 - 7y$
   a. commutative
   b. associative
   c. distributive

   $9n \cdot m = 9mn$
   a. commutative
   b. associative
   c. distributive

Factor each expression using the distributive property.

2. $25a - 5$ $\qquad$ $40x + 12y$ $\qquad$ $-9m - 27n$

3. $21y + 6xy$ $\qquad$ $a - 8ab$ $\qquad$ $-16m + 4n - 12$

Simplify each expression by canceling.

4. $\dfrac{7xy}{28x}$ $\qquad$ $-\dfrac{5}{12} \cdot \dfrac{3}{25}$ $\qquad$ $\dfrac{2b}{9} \cdot \dfrac{18}{b}$

5. $\dfrac{16m}{3n} \cdot -\dfrac{9}{4m}$ $\qquad$ $\dfrac{3b}{4} \cdot \dfrac{20a}{9} \cdot \dfrac{1}{2ab}$ $\qquad$ $-\dfrac{x}{yz} \cdot \dfrac{27y}{14} \cdot \dfrac{7xz}{9x}$

Choose the correct value for each expression.

6. $(2 + 1 \cdot 4)^2 - 6$
   a. 6
   b. 18
   c. 30

   $10 \cdot 4 - 2 \cdot 3^2$
   a. 4
   b. 22
   c. 28

   $\dfrac{48}{4 + 2 \cdot 2}$
   a. 6
   b. 12
   c. 16

Find the value of each expression.

7. $9 + 4(6 - 5)^2$ $\qquad$ $\dfrac{(11 - 3 \cdot 2)^2}{5}$ $\qquad$ $\left[\dfrac{7(24 \div 6) - 4 \cdot 3}{2}\right]^2$

Solve.

8. Jared played in five games of a basketball tournament. The points he scored in each game are as follows: 17, 23, 30, 16, and 24. Write an expression to represent Jared's average points per game for the tournament. Then find the value of the expression.

Simplify using positive exponents.

9. $5xy^3 \cdot x^2y$ $(2a^{-2}b^2c)^3$ $(-3)^{-2}m^3n^0$

10. $\dfrac{12x^8}{6x^5}$ $\dfrac{(2mn)^2}{m^3n}$ $(a^{-2}\,b^1c^3)^{-2}$

Write each expression using only positive exponents.

11. $\dfrac{x^2}{x^5}$ $\dfrac{2m^2n^{-4}}{2^2m^3n^2}$ $\dfrac{a^{-3}b^5}{a^3b^{-2}c^{-2}}$

Simplify the expressions using scientific notation.

12. $(1.2 \times 10^4)(4.0 \times 10^{-6})$ $\dfrac{9.6 \times 10^5}{3.0 \times 10^2}$

Simplify each expression.

13. $3\sqrt{45}$ $\sqrt{128}$ $\sqrt[3]{192}$

14. $\sqrt{\dfrac{9}{25}}$ $\sqrt{\dfrac{200}{18}}$ $\sqrt{\dfrac{5}{27}}$

Find and correct each error.

15. $\sqrt{80x^6} = \sqrt{16 \cdot x^2 \cdot 5x^3} = 4x\sqrt{5x^3}$ $3\sqrt{72m^3} = 3\sqrt{36 \cdot m^2 \cdot 2m} = 39m^2\sqrt{2m}$

Simplify by rewriting with fractional exponents.

16. $\sqrt{x} \cdot \sqrt[5]{x}$ $\left(\sqrt[4]{a^3}\right)^4$

VARIABLES

Algebra is written with numbers, variables, and symbols. A **variable** is a letter used to represent an unknown number. You might think of it as a placeholder for a number. Using variables, you can write expressions that tell the operations you want to perform.

Using Variables to Create Expressions

In this lesson, you will learn to

- Translate words into algebraic symbols.

The first step in writing expressions is to understand the vocabulary. Starting with addition and subtraction, you need to recognize all the ways that a problem can tell you to add or subtract. Study the vocabulary and examples in the chart.

ADDITION		SUBTRACTION	
The *sum* of x and 3	$x + 3$	The *difference* of a and 6	$a - 6$
Length (l) *plus* width (w)	$l + w$	12 *minus* n	$12 - n$
a *increased by* 5	$a + 5$	b *decreased by* c	$b - c$
x *more than* 4	$4 + x$	5 *fewer than* m	$m - 5$
		10 *less than* x	$x - 10$

You know from the commutative property that order doesn't matter in addition. The sum of x and 3 can be written $x + 3$ or $3 + x$. Order does matter in subtraction: $a - 6$ may not equal $6 - a$.

Study the last two subtraction examples carefully. The phrase 10 *less than* x does NOT equal $10 - x$. If you are going to pay $10 less than the sale price of an item, you would subtract $10 from the sale price. It wouldn't make sense to subtract the sale price from $10. For the phrases less than and fewer than, the amount to subtract is stated first and the amount to subtract from is stated second.

Now look at multiplication and division.

> **Think About It:**
> Can you think of a value for a that makes $a - 6$ equal to $6 - a$?

MULTIPLICATION		DIVISION	
The *product* of a and b	ab	The *quotient* of 12 and y	$\frac{12}{y}$
Length (l) *times* width (w)	$l \cdot w$	12 *divided by* y	$\frac{12}{y}$
Twice the radius (r)	$2r$		

(**Note:** Although order doesn't matter in multiplication, it is common practice to write numbers before variables. So, the product of n and 5 is $5n$.)

(**Note:** In algebra, division is shown as a fraction.)

Below are some examples that combine several operations in one expression.

IN WORDS	AS ALGEBRA	EXPLANATION
Four times the sum of x and y	$4(x + y)$	Parentheses show that the sum is a single quantity.
The sum of 7 times a and b	$7a + b$	The word *and* helps you decide where to put the + sign.
Twice the difference of l and w	$2(l - w)$	The difference is a single quantity that you are doubling.
The quotient of the quantity x plus y and 4	$\dfrac{x + y}{4}$	The sum is a single quantity, but parentheses are not needed because the fraction bar acts as a grouping symbol.
The product of 5 and n squared less than 3	$3 - 5n^2$	Parentheses are not needed because multiplication comes before subtraction in the order of operations.

Write each phrase as an algebraic expression.

1. *m* decreased by *n*   5 more than *b*

2. the product of *x*, *y*, and 3   8 less than the quotient of *a* and 2

3. the difference of *x* squared and the product of *x* and *y*   the difference of *b* and *c*, divided by the *sum of* *b* and *c*

4. three times the difference of *m* and *n*   the product of 4 and the sum of *x* squared and *y*

Write each phrase as an algebraic expression. Let *x* represent the unknown number.

5. the product of 3 more than a number and 3 less than the same number   6 times the difference of a number and 7

6. the quotient when 5 less than a number is divided by the number to the third power   the product of 5 and a number, increased by that same number

7. half of the quantity of a number increased by 9 (**Hint:** *half of* means dividing by 2 or multiplying by $\frac{1}{2}$.)   the product of the sum of a number and 5 and the difference of that number and 5

Simplifying Expressions

In this lesson, you will learn to

- Combine like terms.
- Add and subtract polynomials.
- Multiply polynomials.

The building blocks of an expression are terms. In math, a **term** is defined as a number or the product of a number and one or more variables raised to a power. In an expression, terms are connected by addition and subtraction symbols.

The expression $3x^2 - 4x + 5$ has three terms: $3x^2$, $-4x$, and 5.

You may be wondering whether the symbol (−) is a subtraction symbol or a negative sign. In a way, it's both. You can think of subtraction as a direction to "add the opposite," so $3x^2 - 4x$ equals $3x^2 + (-4x)$. You can also think of a sign (whether + or −) as attached to the term that follows it.

Just as you can add only things that are alike, you can combine only like terms. **Like terms** have exactly the same variables raised to exactly the same powers.

$5xy^2$ and $-2xy^2$ are like terms. $\qquad$ $3x^2y$ and $4xy^2$ are not like terms.

To add or subtract like terms, simply combine the numbers, keeping the variables and powers the same.

EXAMPLE 1 $\quad$ Simplify: $3a^2 + 7 - 2b + 3 - 4a^2$

STEP 1 Change the order and group the like terms.

$$(3a^2 - 4a^2) + (7 + 3) - 2b$$
$$= \quad -a^2 \quad + \quad 10 \quad - 2b$$
$$= -a^2 + 10 - 2b$$

STEP 2 Combine the like terms.

> When the number before the variable is 1 or −1, leave it out.
> $3a^2 - 4a^2 = -1a^2$, so write $-a^2$.

Simplify each expression by combining like terms.

1. $6x - 4 + 3x + 6$ $\qquad\qquad\qquad$ $-2a - 8b - 5a + 10b$

2. $12m - 13 + 3m + 5n$ $\qquad\qquad\qquad$ $7x^2 - 9 + 4x^2 + 5x - 6$

3. $-5x^3 + x^2 + 2x^3 - 2x^2$ $\qquad\qquad\qquad$ $4y^2 - 3y + 5 - 4y^2 - 6$

4. $14cd - 11c^3 + 7d^3 - 5c^3 - 6cd$ $\qquad\qquad$ $-7m^2n + 4mn^2 - 6mn + 9m^2n - 2mn + 3m^2n$

A **polynomial,** the most common kind of algebraic expression, is a sum of terms. The terms in a polynomial must use only whole number exponents (0, 1, 2, 3, . . .).

Some special names apply to polynomials. A **monomial** has one term, a **binomial** has two terms, and a **trinomial** has three terms. There is no reason to name polynomials with more than three terms.

Mathematicians write polynomials in a certain order known as **standard form.** The terms are written so that the exponents descend, or get smaller, going from left to right. A number without a variable (also called a **constant**) goes at the end.

EXAMPLE 2   Write $5 + a + a^3 + 2a^2$ in standard form. $a^3 + 2a^2 + a + 5$

To add and subtract polynomials, combine like terms.

EXAMPLE 3   Add $4x^2 + 3x - 2$ and $-2x^2 + 5$.

STEP 1 Change the order and group like terms. This is based on the commutative and associative properties. You can probably do this mentally as you work.

$$(4x^2 + 3x - 2) + (-2x^2 + 5)$$
$$= (4x^2 - 2x^2) + 3x + (-2 + 5)$$
$$= 2x^2 + 3x + 3$$

STEP 2 Add the like terms.

To subtract, remember that subtraction is adding the opposite. To find the opposite of an expression, change the sign of every term in the expression. For example, the opposite of $3x - 4$ is $-3x + 4$.

EXAMPLE 4   Subtract: $(5a^2 + 2) - (-2a^2 + 1)$

STEP 1 Because you are subtracting $(-2a^2 + 1)$, rewrite the expression so that you are adding the opposite: $(2a^2 - 1)$.

STEP 2 Group like terms.

$$(5a^2 + 2) - (-2a^2 + 1)$$
$$= (5a^2 + 2) + (2a^2 - 1)$$
$$= (5a^2 + 2a^2) + (2 - 1)$$
$$= 7a^2 + 1$$

STEP 3 Add.

Both addition and subtraction can be done by writing like terms in columns. Many students find that they make fewer mistakes using the column method. In the next example, notice that the first step groups like terms by putting them in a column.

EXAMPLE 5   Subtract using columns: $(5n^3 + 3n + 2) - (2n^3 - 7)$

STEP 1 Arrange the terms in columns. Line up terms with the same variables and powers.

$$\begin{array}{r} 5n^3 + 3n + 2 \\ - (2n^3 \qquad - 7) \\ \hline \end{array}$$

STEP 2 Change subtraction to adding the opposite by changing the sign of every subtracted term.

$$\begin{array}{r} 5n^3 + 3n + 2 \\ + (-2n^3 \qquad + 7) \\ \hline \end{array}$$

STEP 3 Add.

$$3n^3 + 3n + 9$$

Perform the indicated operations. Watch for addition and subtraction signs between the expressions in parentheses.

5. $(7a + 6) + (4a - 9)$ $(x^2 - 7x + 2) + (2x^2 + 7x - 5)$

6. $(3n^3 + 8) + (n^2 - 5n)$ $(4x^2 + 5x - 10) + (-x^2 - 3x + 7)$

7. $(c^3 + 2c^2 - 4c + 2) + (2c^3 + 5c - 1)$ $(2y - 5) - (y + 4)$

8. $(-3n^2 + 4n) - (n^2 + 5n)$ $(4a^2 - a + 6) - (-2a^2 - 3a + 8)$

9. $(2x - 5) - (-4x^2 - 2x + 1)$ $(7m - 2n) + (8m + n) - (10m - 3n)$

(**Hint:** Perform the addition first, and then do the subtraction.)

To multiply polynomials, you will use the rules of exponents and the distributive property.

EXAMPLE 6 Multiply: $4x^2y \cdot 3xy^2$

STEP 1 Show multiplication by writing both terms in parentheses next to each other.

STEP 2 Change the order and group the variables.

STEP 3 Multiply the numbers and add the exponents.

$$(4x^2y)(3xy^2)$$
$$= (4 \cdot 3)(x^2 \cdot x)(y \cdot y^2)$$
$$= 12x^3y^3$$

With a little practice, you will be able to do multiplication steps in your head. This next example shows multiplying a binomial by a monomial. Notice how the distributive property makes this possible.

EXAMPLE 7 Multiply: $3a(a + 5)$

STEP 1 Multiply each term in parentheses by $3a$.

STEP 2 Combine.

$$3a(a + 5) = 3a \cdot a + 3a \cdot 5$$
$$= 3a^2 + 15a$$

As you study the next example, pay attention to the subtraction symbol. Think of the minus sign as a negative sign attached to the term that follows it.

EXAMPLE 8 Multiply: $2x(x^2 - 3x + 4)$

Multiply each term in parentheses by $2x$.
Think: $2x \cdot x^2$, $2x \cdot -3x$, and $2x \cdot 4$

$$2x(x^2 - 3x + 4)$$
$$= 2x^3 - 6x^2 + 8x$$

You will need to know how to multiply monomials to solve many algebra problems. Remember, a binomial has two terms. To multiply binomials, make sure that every term in the first factor is multiplied by every term in the second factor.

You may have learned the FOIL method in the past. FOIL stands for First-Outer-Inner-Last. Study how it works in the next example.

EXAMPLE 9   Multiply: $(x + 4)(x - 5)$

F First   Find the product of the first terms of the factors. $x \cdot x = x^2$
$(\mathbf{x} + 4)(\mathbf{x} - 5)$

O Outer   Find the product of the outer terms. $(\mathbf{x} + 4)(x - \mathbf{5})$ $x \cdot -5 = -5x$
Think of the minus sign as traveling with the 5.

I Inner   Find the product of the inner terms. $(x + \mathbf{4})(\mathbf{x} - 5)$ $4 \cdot x = 4x$

L Last   Find the product of the last terms of the factors. $4 \cdot -5 = -20$
$(x + \mathbf{4})(x - \mathbf{5})$

Combine the results of each step: $x^2 - 5x + 4x - 20$

Simplify by adding like terms: $x^2 - x - 20$

Multiply. If necessary, simplify by combining like terms.

10. $x^2y(5xy)$	$2ab^2(-6a^4b^3)$	$2n^2(-3m^5n)$
11. $n(2m + 9)$	$8a(a + 3b)$	$-y(4x - y)$
12. $11x(-x + 2y)$	$-7y(3x + 2y)$	$5m(m - 6n)$
13. $-2(m^2 + 7m - 5)$	$3a(a^2 + 5a + 15)$	$-b(2b^2 - 3b - 1)$
14. $4x(x^2 - 3x - 6)$	$-2y(y^2 + y - 8)$	$5n(3n^2 + 7n - 2)$
15. $(m + 4)(m + 3)$	$(a - 5)(a - 2)$	$(2c + 1)(c - 6)$
16. $(m + 5)(m - 5)$	$(y + 8)(y + 2)$	$(x - 9)(-x + 2)$
17. $(3y - 4)(2y + 2)$	$(1 - a)(5 - a)$	$(2x + 3y)(x - 2y)$
18. $(x + 4)^2$	$(n - 5)^2$	$(5 - a)(a + 3)$
(**Hint:** Write it out as $(x + 4)(x + 4)$.)		(**Hint:** Terms don't have to be in the same order to use FOIL.)

Working with Fractions in Expressions

In this lesson, you will learn to

- Add and subtract like and unlike fractions in algebraic expressions.

- Simplify fractional expressions by factoring.

Think back to basic math. You know that you can add or subtract fractions only when they have a common, or the same, denominator. In algebra, the same rule applies.

Follow these steps for adding or subtracting fractions with like denominators.

STEP 1 Add or subtract the numerators.

STEP 2 Write the sum or difference over the common denominator.

STEP 3 Simplify by canceling if necessary.

EXAMPLE 1 Add: $\dfrac{9}{2x} + \dfrac{3}{2x}$

Add the numerators:

$$\frac{9+3}{2x} = \frac{12}{2x}$$

Simplify:

$$\frac{\overset{6}{\cancel{12}}}{\underset{1}{\cancel{2x}}} = \frac{6}{x}$$

> **Think About It:** You can simplify because both the numerator and the denominator can be divided by 2. Can you explain why this works?

When polynomials are involved, the process is still the same.

EXAMPLE 2 Subtract: $\dfrac{5y}{y-6} - \dfrac{2}{y-6}$

Set up the problem and subtract the numerators.

$$\frac{5y}{y-6} - \frac{2}{y-6} = \frac{5y-2}{y-6}$$

This expression cannot be simplified further.

Check to see if the numerator or the denominator of a fraction can be factored. Factoring makes expressions easier to manage and may lead to an opportunity to cancel.

EXAMPLE 3 Add: $\dfrac{3x+8}{x+4} + \dfrac{4}{x+4}$

Add the numerators:

$$\frac{3x+8}{x+4} + \frac{4}{x+4} = \frac{3x+12}{x+4}$$

Factor the numerator:

$$\frac{3x+12}{x+4} = \frac{3(x+4)}{x+4}$$

Simplify by canceling.

$$\frac{3\cancel{(x+4)}}{\cancel{x+4}} = 3$$

Add or subtract. Then simplify.

1. $\dfrac{7x}{16} - \dfrac{5x}{16}$

     $\dfrac{2m}{n} - \dfrac{5m}{n}$

2. $\dfrac{2c}{c-3} - \dfrac{3}{c-3}$

     $\dfrac{3x-8}{x-6} + \dfrac{x-16}{x-6}$

3. $\dfrac{9}{2m} - \dfrac{5}{2m}$

     $\dfrac{n-1}{3} + \dfrac{2n-8}{3}$

To add or subtract unlike fractions, you must first rewrite the fractions so they have like denominators. Review the process using ordinary fractions first.

Suppose you need to add $\frac{1}{2}$ and $\frac{1}{3}$. Find the lowest common denominator (LCD). The LCD is the lowest number that is divisible by both denominators. The LCD for $\frac{1}{2}$ and $\frac{1}{3}$ is 6. Rewrite the fractions so they both have 6 as a denominator. Remember that you must multiply the numerator and the denominator of a fraction by the same number.

$$\frac{1}{2} \cdot \frac{3}{3} = \frac{3}{6} \text{ and } \frac{1}{3} \cdot \frac{2}{2} = \frac{2}{6}$$

 Think About It: Multiplying by $\frac{3}{3}$ or $\frac{2}{2}$ doesn't actually change the value of the fraction. Why not?

Now you can add the fractions: $\frac{3}{6} + \frac{2}{6} = \frac{5}{6}$. The answer is already in simplest form.

Try an example from algebra.

EXAMPLE 4 Add: $\frac{2}{3x} + \frac{1}{x^3}$

STEP 1 To find the LCD, factor both denominators completely. The new denominator will contain each factor the greatest number of times it appears in a single denominator.

$3x = 3 \cdot x$
$x^3 = x \cdot x \cdot x$

The LCD must contain 3 and x, but x must appear three times because it appears three times in the second denominator.

The LCD is $3 \cdot x \cdot x \cdot x$, or $3x^3$.

STEP 2 Rewrite each fraction with the LCD.

$$\frac{2}{3x} \cdot \frac{x^2}{x^2} = \frac{2x^2}{3x^3} \qquad \frac{1}{x^3} \cdot \frac{3}{3} = \frac{3}{3x^3}$$

STEP 3 Add the like fractions.

$$\frac{2x^2}{3x^3} + \frac{3}{3x^3} = \frac{2x^2 + 3}{3x^3}$$

Think About It: By what number can I multiply to make this denominator equal the LCD? Then multiply both the numerator and the denominator by that quantity.

Add or subtract. Then simplify.

4. $\frac{3}{a} + \frac{5}{a^2}$

$\frac{3}{x} - \frac{2}{y}$

5. $\frac{2}{5x^2} + \frac{3}{4x^3}$

$\frac{1}{n^2} - \frac{2}{5n}$

6. $\frac{6}{xy} + \frac{3}{y^2}$

$\frac{4a}{bc} - \frac{2c}{ab}$

Follow the same steps when there are binomials in the denominator. Often the LCD will be the product of the two denominators.

EXAMPLE 5 Subtract and simplify: $\dfrac{3}{m+2} - \dfrac{2}{m}$

STEP 1 The LCD must have the factors m and $m + 2$. Multiply the factors. The LCD is $m(m + 2)$.

STEP 2 Rewrite each fraction. Notice that in the second fraction, the distributive property is applied to complete the multiplication in the numerator.

$$\frac{3}{m+2} \cdot \frac{m}{m} = \frac{3m}{m(m+2)} \qquad \frac{2}{m} \cdot \frac{m+2}{m+2} = \frac{2(m+2)}{m(m+2)} = \frac{2m+4}{m(m+2)}$$

STEP 3 Combine the like fractions:

Parentheses are needed to show the entire expression is subtracted.

$$\frac{3m}{m(m+2)} - \frac{2m+4}{m(m+2)} = \frac{3m - (2m+4)}{m(m+2)} = \frac{3m - 2m - 4}{m(m+2)} = \frac{m-4}{m(m+2)}$$

Sometimes two denominators are opposites of each other. Multiply one of the fractions by $\dfrac{-1}{-1}$ to make the denominators the same.

EXAMPLE 6 Simplify: $\dfrac{x^2}{x-4} + \dfrac{3}{4-x}$

STEP 1 The LCD $(x - 4)(4 - x)$ could be used to solve the problem, but there is an easier way. Multiply the second fraction by $\dfrac{-1}{-1}$.

$$\frac{3}{4-x} \cdot \frac{-1}{-1} = \frac{-3}{-4+x} = \frac{-3}{x-4}$$

STEP 2 Now you have two like fractions. Add.

$$\frac{x^2}{x-4} + \frac{-3}{x-4} = \frac{x^2-3}{x-4}$$

Remember to look for opportunities to factor the denominator.

EXAMPLE 7 Subtract and simplify: $\dfrac{4}{3m-6} - \dfrac{1}{2m-4}$

STEP 1 To find the LCD, first factor both denominators.

$$\frac{4}{3m-6} = \frac{4}{3(m-2)} \qquad \frac{1}{2m-4} = \frac{1}{2(m-2)}$$

Rewrite the problem: $\dfrac{4}{3(m-2)} - \dfrac{1}{2(m-2)}$

The LCD is $3 \cdot 2 \cdot (m-2) = 6(m-2)$.

STEP 2 Use the LCD to rewrite the fractions.

$$\frac{4}{3(m-2)} \cdot \frac{2}{2} = \frac{8}{6(m-2)} \text{ and } \frac{1}{2(m-2)} \cdot \frac{3}{3} = \frac{3}{6(m-2)}$$

STEP 3 Subtract. $\dfrac{8}{6(m-2)} - \dfrac{3}{6(m-2)} = \dfrac{5}{6(m-2)} \text{ or } \dfrac{5}{6m-12}$

In the last example, there were two ways to write the solution. Both are correct, but you should generally leave the denominator factored, as this is useful in more complex problems.

Add or subtract. Factor numerators and denominators.

7. $\dfrac{y}{y+1} + \dfrac{3}{5}$ $\qquad\qquad\qquad\qquad\qquad$ $\dfrac{5}{6} - \dfrac{m}{m+1}$

8. $\dfrac{7}{b-1} + \dfrac{3}{b}$ $\qquad\qquad\qquad\qquad\qquad$ $\dfrac{4}{3x-1} + \dfrac{2}{3x}$

9. $\dfrac{1}{c-2} + \dfrac{4}{3c-6}$ $\qquad\qquad\qquad\qquad$ $\dfrac{2}{5y+5} - \dfrac{5}{2y+2}$

10. $\dfrac{1}{b-2} + \dfrac{4}{5b-10}$ $\qquad\qquad\qquad\qquad$ $\dfrac{5}{4-n} + \dfrac{2}{n-4}$

11. $\dfrac{3}{2-a} - \dfrac{1}{a-2}$ $\qquad\qquad\qquad\qquad\quad$ $\dfrac{5}{x-7} + \dfrac{2}{7-x}$

12. $\dfrac{4}{a+5} + \dfrac{1}{a-1}$ $\qquad\qquad\qquad\qquad$ $\dfrac{4}{3x-9} + \dfrac{3x}{2x-6}$

    (**Hint**: The LCD is $(a+5)(a-1)$.)

13. $\dfrac{6}{y+4} - \dfrac{3}{y-8}$ $\qquad\qquad\qquad\qquad$ $\dfrac{4}{a+3} + \dfrac{2}{a-1}$

14. $\dfrac{c}{(c-1)(c+1)} + \dfrac{2}{c+1}$ $\qquad\qquad$ $\dfrac{5}{x-4} + \dfrac{2}{(x+1)(x-4)}$

15. **Challenge** $\dfrac{4a}{(a-1)(a-2)} - \dfrac{1}{(a-2)}$ $\qquad$ $\dfrac{2}{(x+2)(x+3)} - \dfrac{3x}{(x-5)(x+3)}$

Evaluating Expressions

In this lesson, you will learn to

- Find the value of algebraic expressions.

- Evaluate formulas.

To *evaluate an expression* means to find the value of the expression. When an expression is written with variables, you must first substitute the values you are given for the variables. Then use the order of operations to evaluate the expression.

Consider the expression $4x$. The expression states that some number x is multiplied by 4. If the value of x is -2, the value of the expression is $4(-2) = -8$. If the value of x is $+2$, the value of the expression is $4(2) = 8$. This illustrates an important point. The value of a variable can be positive or negative.

The next example shows the steps for evaluating any expression.

EXAMPLE 1 Evaluate $2x + 5y$ if $x = -3$ and $y = 2$.

STEP 1 Replace the variables with the given values.

STEP 2 Complete the operations following the order of operations.

$$2x + 5y = 2(-3) + 5(2)$$
$$= -6 + 10$$
$$= 4$$

Subtraction and negative signs sometimes cause confusion. Study this next example carefully to see how to handle these signs.

EXAMPLE 2 Evaluate $7 - 2b$ if $b = -5$.

STEP 1 Replace the variable with the value.

STEP 2 Do the multiplication: $2(-5) = -10$.

STEP 3 This step involves subtracting a negative. Think: *Two negatives make a positive.*

STEP 4 Add.

$$7 - 2b = 7 - 2(-5)$$
$$= 7 - (-10)$$
$$= 7 + 10$$
$$= 17$$

Take another look at Step 3: $7 - (-10) = 7 + 10$. Why did that work? Subtraction means "adding the opposite," so instead of subtracting -10, you add the opposite of -10, which is $+10$.

In some expressions a sum or difference is being subtracted.

EXAMPLE 3 Evaluate $a - (b - c)$ if $a = 5$, $b = -2$, and $c = -6$.

STEP 1 Replace the variables with the given values.

STEP 2 Do the work in parentheses. The two negatives become a positive: $(-2 - -6)$ becomes $(-2 + 6)$.

STEP 3 Compute the addition in parentheses.

STEP 4 Subtract.

$$a - (b - c)$$
$$= 5 - (-2 - -6)$$
$$= 5 - (-2 + 6)$$
$$= 5 - 4$$
$$= 1$$

Evaluate the expressions if $a = -3$, $b = 5$, $c = -1$, and $d = 4$.

1. $5a + 3d$ $4b - 3a$ $a - (b - c)$

2. $2a - (b + c)$ $a(c - d)$ $3b - (4c + 2a)$

The next example includes an exponent. Remember to evaluate exponents before you multiply, divide, add, or subtract.

EXAMPLE 4 Evaluate $2 - m^2$ if $m = -3$.

STEP 1 Replace the variable with given value.

STEP 2 Evaluate the exponent: $(-3)^2 = 9$.

STEP 3 Subtract.

$$2 - m^2 = 2 - (-3)^2$$
$$= 2 - 9$$
$$= -7$$

Think About It:
Why don't the two negative signs become a positive sign in this example?

Finally, take a look at a division example. As you work with division, remember that the fraction bar works like parentheses. Always perform the work in the numerator and the denominator separately before you perform the final division step.

EXAMPLE 5 Evaluate $\dfrac{5(x + y)}{2x + y}$ if $x = 3$ and $y = -1$.

STEP 1 Replace the variables with the given values.

STEP 2 Evaluate the top and bottom expressions.

STEP 3 Divide.

$$\frac{5(x + y)}{2x + y} = \frac{5(3 + -1)}{2(3) + (-1)}$$
$$= \frac{5(2)}{6 + -1}$$
$$= \frac{10}{5} = 2$$

Evaluate the expressions if $a = -2$, $b = 3$, $c = -6$, $d = 4$, $e = -1$, and $f = 2$.

3. $\dfrac{-5c}{b}$ $\dfrac{3f - 2a}{d - e}$ $\dfrac{4c}{be}$ $\dfrac{(ac)^2}{bd}$

4. $c^2 - e^2$ $-b^2 - (a + d)$ $-(e + f)^2$ $\dfrac{3(c + d)(c - d)}{10b}$

5. $\dfrac{a + b - c}{e^2}$ $\dfrac{4d - 2a}{d - e}$ $e^3 + a^3$ $\dfrac{a^2 + d^2}{-f}$

Synthesis   A **formula** uses algebra to express a procedure for solving a problem. The formulas for finding the perimeter and area of four basic shapes are shown below.

Remember, **perimeter** is the distance around a figure, and **area** is the measure of the surface area inside a figure.

Area is always measured in square units.

Square

Perimeter = 4s

Area = s^2

Rectangle

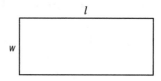

Perimeter = 2*l* + 2*w*

Area = *lw*

In formulas, letters are used to represent parts of the figure.

s = side

l = length

w = width

b = base

h = height

Triangle

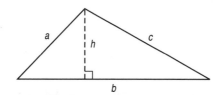

Perimeter = *a* + *b* + *c*

Area = $\frac{1}{2}bh$ or $\frac{bh}{2}$

Trapezoid

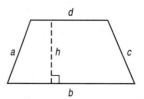

Perimeter = *a* + *b* + *c* + *d*

Area = $h\left(\dfrac{b + d}{2}\right)$

Use the formulas above to solve the problems.

6. The side of a square measures 11 in.   **a.** Find the perimeter.   **b.** Find the area in square inches.

7. A rectangle has a length of 16 cm and a width of 12 cm.   **a.** Find the perimeter.   **b.** Find the area in square centimeters.

8. 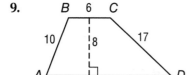   **a.** Find the perimeter of triangle *XYZ*.   **b.** Find the area of △*XYZ* in square units.

9. *B* 6 *C*, 10, 8, 17, *A*, 18, *D*   **a.** Find the perimeter of trapezoid *ABCD*.   **b.** Find the area of trapezoid *ABCD* in square units.

Variables Review

Solve the problems below. When you finish, check your answers at the back of the book and correct any errors.

Write an algebraic expression for each phrase.

1. 13 increased by the product of 5 and m

   x less than y

2. the quotient of 10 and n

   3 times the difference of P and Q

3. the sum of b and c, divided by the product of b and c

   the difference of the square of m and n

Write each phrase as an algebraic expression. Let x represent the unknown number.

4. twice the sum of a number and 8

   the product of a number and 5, divided by the number raised to the third power

Simplify each expression by combining like terms.

5. $4x - 5 + 3x + 8$ $-13n - 9m + 5m + 20n$

6. $-3a^2 + 12 - 7a^2 - 3$ $5y^2 - 5 - 3y - y^2 + 8$

Perform the indicated operations and simplify.

7. $(4x^3 + 3) + (8 - 3x^3)$ $(3a - 7) - (a + 5)$

8. $(n^2 - 5n + 6) + (3n^2 + 4n - 7)$ $(2x^2 - x + 5) - (3x^2 + 5x - 4)$

9. $(4n - 3) - (3n^2 + 5n + 9)$ $(3a - 2b) + (4a - b) - (5a + 6b)$

Multiply. Then simplify if necessary.

10. $a^3c \cdot 4abc^2$ $6x(x + 2y)$ $-7n(6 + 5n)$

11. $4a(3a + b)$ $2x^3(x - y)$ $(d - 1)(d + 2)$

12. $(5a - 3)(a + 3)$ $(m + 2)(7 - m)$ $3x^2y^2(x + y)$

13. $(b + 4)(b - 3)$ $(2d + 5)(d - 3)$ $(x - 4)^2$

14. $(3 - m)(2 + m)$ $(x - 7)(5 - x)$ $(n + 3)^2$

Add or subtract. Then simplify.

15. $\dfrac{x - 1}{4} + \dfrac{3x + 5}{4} + \dfrac{2x - 9}{4}$ $\dfrac{3y - 4}{y - 2} - \dfrac{5y + 3}{y - 2}$

16. $\dfrac{5}{x} - \dfrac{2}{y}$ $\dfrac{3}{a^2} - \dfrac{4}{a}$

17. $\dfrac{4}{n + 1} + \dfrac{1}{n}$ $\dfrac{y}{y - 1} + \dfrac{2}{1 - y}$

18. $\dfrac{3}{2m + 6} - \dfrac{2}{4m + 12}$ $\dfrac{x}{(x + 2)(x - 3)} - \dfrac{1}{(x - 3)}$

Evaluate the expressions. Let $a = -2, b = 6, c = -3,$ and $d = 1$.

19. $2a(b - c)$ $\qquad\qquad\qquad$ $c - (d - b)$ $\qquad\qquad\qquad$ $\dfrac{3b}{ac}$

20. $a^2 + b^2$ $\qquad\qquad\qquad$ $\dfrac{(-b)^2}{-ac}$ $\qquad\qquad\qquad$ $(ab)^2 - cd$

Evaluate the expressions. Let $w = -1, x = 4, y = -5,$ and $z = 2$.

21. $-y^2 + wz$ $\qquad\qquad\qquad$ $\dfrac{-2xy}{z^2}$ $\qquad\qquad\qquad$ $\dfrac{-yz}{x - w}$

22. $\dfrac{y - z}{w - x - z}$ $\qquad\qquad\qquad$ $x^3 - y^2$ $\qquad\qquad\qquad$ $y^2 - 2z^3 + w$

Use the formulas on page 27 to solve the following problems.

23. A rectangle has a length of 15 feet and a width of 8 feet. $\qquad$ **a.** Find the perimeter. $\qquad$ **b.** Find the area in square feet.

24. The side of a square measures 14 meters. $\qquad$ **a.** Find the perimeter. $\qquad$ **b.** Find the area in square meters.

25. 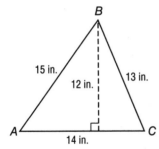 $\qquad$ **a.** Find the perimeter of $\triangle ABC$ in inches. $\qquad$ **b.** Find the area of $\triangle ABC$ in square inches.

26. 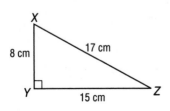 $\qquad$ **a.** Find the perimeter of $\triangle XYZ$ in centimeters. $\qquad$ **b.** Find the area of $\triangle XYZ$ in square centimeters.

EQUATION AND INEQUALITY BASICS

In the last section, you worked with expressions, the building blocks of an algebraic sentence. Now you're ready to put the pieces together. Mathematical statements, especially equations, are some of the most powerful tools in the language of algebra.

Beyond Expressions

In this lesson, you will learn to

- Graph the solution to a one-variable equation or inequality.
- Decide whether a given number is a solution for an equation or inequality.

An **equation** is a mathematical statement that two expressions are equal. Formulas are examples of equations. For example, you know that the area of a rectangle is equal to length times width: $A = lw$.

An equal sign (=) separates the left and right sides of an equation.

EXAMPLES $2 + 5 = 7$ $A = \frac{1}{2}bh$ $x = 3$

Equations, just like sentences, can be true or false. When you solve an equation, you are looking for values of the variables that will make the equation true.

Take another look at the equation $x = 3$. This sentence is only true if x is 3. In other words, out of the entire set of real numbers, only 3 satisfies the conditions of the equation.

You can graph the solution on a number line.

Graphing offers a quick way to view number relationships. Graphing is very helpful with inequalities and equations with more than one variable.

Some equations have more than one solution, or **root.** For example, $x^2 = 16$ has two possible solutions: **4** and **−4.** This equation is an example of a quadratic equation, which is covered later in this book.

An **inequality** is a mathematical statement that shows the relationship between two non-equal expressions. The chart below shows the inequality symbols.

SYMBOL	MEANING	EXAMPLE
<	less than	$2 < 7$
≤	less than or equal to	$3 \leq 4$ or $3 \leq 3$
>	greater than	$8 > 3$
≥	greater than or equal to	$7 \geq 5$ or $7 \geq 7$
≠	not equal to	$5 \neq 8$

Hint: Many students think of the symbols < and > as arrows that point to the smaller value.

Inequalities can also be true or false. Creating a graph of an inequality makes it much easier to determine whether the solution to an inequality is true.

EXAMPLE 1 Graph $x < 2$.

Any value less than 2 will make the inequality true. Circle 2 on the number line, and then fill in the line to the left of 2.

The open circle on the number 2 shows that 2 is *not* included in the solution set of this inequality. In other words, the value 2 does *not* make the inequality true.

EXAMPLE 2 Graph $a \geq -3$.

The symbol $\geq$ means *greater than or equal to,* so the value -3 is part of the solution set. Put a closed circle at -3, and fill in the line to the right.

Remember: On a graph, a closed (filled-in) circle means the value is included in the solution set, while an open circle means the value is not a solution.

Graph the solutions for the equations and inequalities. Add points and markings to the graph as needed.

1. $x = -2$

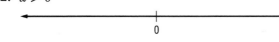

   $y < 3$

2. $a > 0$

   $b \leq -1$

3. $n = 4$

   $m \geq 1$

4. $c > -3$

   Challenge $d \neq 1$

Whether equations and inequalities are true or false depends on the value of the variable. If a given value of a variable makes the statement true, then that value is a solution of the statement.

EXAMPLE 3 For the equation $5x - 3 = 2x + 3$, is 2 a solution?

STEP 1 Substitute 2 for x.

STEP 2 Evaluate both sides of the equation using the order of operations.

$$5x - 3 = 2x + 3$$
$$5(2) - 3 = 2(2) + 3$$
$$10 - 3 = 4 + 3$$
$$7 = 7 \text{ True}$$

STEP 3 Examine the result. Since $7 = 7$ is a true statement, the value **2 is a solution** for this equation.

You also can substitute a value for the variable in an inequality. Evaluate both sides carefully, and then examine the result.

EXAMPLE 4   For the inequality $6m + 7 < 2m - 15$, is -3 a solution?

STEP 1   Substitute -3 for m.

STEP 2   Evaluate both sides of the inequality using the order of operations.

$$6m + 7 < 2m - 15$$
$$6(-3) + 7 < 2(-3) - 15$$
$$-18 + 7 < -6 - 15$$
$$-11 < -21 \text{ False}$$

STEP 3   Because -11 is greater than -21, the statement is false. The value **-3 is not a solution** for this inequality.

An inequality result that compares two negative numbers can be difficult to evaluate. At first glance, $-11 < -21$ may appear true because 11 is less than 21. Both values are negative, however, so -11 is actually greater than -21.

If you have trouble, think about where you would place the values on a number line in relation to 0.

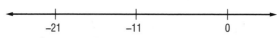

On a number line, values get greater as you move to the right.

Using the number line, you can clearly see that -11 is greater than -21.

Write *True* if the value is a solution for the equation or inequality. Write *False* if it is not a solution. The first problem is done for you.

5. Does $a = 7$? $4a - 7 = 3a$
$$4(7) - 7 = 3(7)$$
$$28 - 7 = 21$$
$$21 = 21 \text{ True}$$

Does $b = -2$? $6b + 7 = 3b - 2$

6. Does $x = -8$? $4x - 9 = 2x + 7$

Does $y = 3$? $5y - 3 = 15 - y$

7. Does $n = 5$? $2n - 9 = 16 - 3n$

Does $a = -1$? $6 - 2a = 3 - a$

8. Could $x = 3$? $6 - x < 4x - 9$

Could $x = 0$? $10x > 4x - 12$

9. Could $b = 3$? $5b - 2 > 6 + b$

Could $n = -5$? $3n + 2 \leq n - 6$

10. Could $x = 1$? $13 - (2x + 2) \leq 2(x + 2) + 3x$

Could $m = -4$? $3(m - 2) \geq 1 - 5(m + 5)$

The Rules of Equality

In this lesson, you will learn to

- Apply the rules of equality to write equivalent equations.

Solving equations is often the goal of algebra. To solve an equation, use strategies and operations to simplify the equation until the answer is clear. As you work, remember this important rule: The equation must remain balanced.

For example, the equation $3x - 5 = 4x - 9$ could be represented by the drawing at the right.

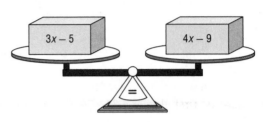

Each side of the equation is balanced by the other. Think of the two sides of the scale as having the same weight. Do not upset that balance as you work with the equations. The **rules of equality** explain what you can and cannot do in order to maintain that balance.

The first two rules may seem obvious, but pay close attention to the ideas. You will need them often as you work with equations.

RULE NAME	IN SYMBOLS	IN WORDS
Symmetric	If $a = b$, then $b = a$.	You can switch the sides of an equation.
Transitive	If $a = b$ and $b = c$, then $a = c$.	You can substitute one equal value for another equal value.

To discover the remaining rules of equality, take another look at a balance scale.

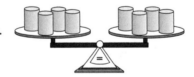

Each side of the scale holds four identical objects. Suppose you wanted to add one more of the same block to each side of the scale. Would the scale still balance? Of course it would.

What if you wanted to double the number of blocks on each side? Would the scale balance? Yes, the weight on each side would change, and there would be more objects, but the scale would still balance.

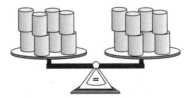

In fact, the scale will remain balanced after you perform any operation (except dividing by 0) as long as you perform the operation on both sides of the scale.

The **additive property of equality** says you can add or subtract the same quantity on both sides of an equation. If $a = b$, then $a + c = b + c$.

The **multiplicative property of equality** says you can multiply or divide by the same *nonzero* number on both sides of an equation. If $a = b$, then $ac = bc$ and $\frac{a}{c} = \frac{b}{c}$.

Do you understand why c cannot be 0? First, the rules of mathematics prevent division by 0; division by zero is undefined. If you multiply both sides by 0, the equation will be $0 = 0$ and the variables will have vanished. This is not useful for finding a solution.

Write equivalent equations by performing the indicated operation on both sides of the equation. Don't solve the equation. The first row has been completed for you.

1. Add 7 to both sides.

$$6a - 7 = 47$$
$$6a - 7 + 7 = 47 + 7$$
$$6a = 54$$

Divide both sides by 3.

$$3x = -15$$
$$\frac{3x}{3} = \frac{-15}{3}$$
$$x = -5$$

2. Multiply both sides by -1.

$$-y = 12$$

Subtract 5 from both sides.
(**Hint:** Subtracting 5 is the same as adding -5.)

$$-9 = 2b + 5$$

3. Add 6 to both sides.

$$-6 - 4n = -5n + 9$$

Multiply both sides by 4.

$$\frac{x}{4} = \frac{9}{2x}$$

4. Divide both sides by -5.

$$-10x = -5(3 + x)$$

Subtract $3a$ from both sides.

$$2a = 3a + 7$$

5. Multiply both sides by $4m$.

$$\frac{3}{4m} = 2$$

Add $7y$ to both sides.

$$21 - 7y = 0$$

6. Subtract x^2 from both sides.

$$x^2 + x + 9 = x^2 + 10$$

Divide both sides by -2.
(**Hint:** Divide each term on the left side.)

$$-6a - 2 = -14$$

Solving an Equation

You have all the rules and tools you need. You are ready to start solving equations.

The goal in solving an equation is to **isolate the variable.** In other words, you want the variable to be alone on one side of the equation, with everything else on the other side.

Imagine that you need to untangle a knotted rope. What would you do to untangle it? First, you would probably study it to find the twist or turn that happened last. This would be the one to untangle first. You would then work backward until every knot was undone. Use this same approach to solve an equation. First study the equation to see what has happened to the variable. Then undo each operation until the variable is isolated.

EXAMPLE 1   Solve: $x - 7 = 3$

$$x - 7 = 3$$
$$x - 7 + 7 = 3 + 7$$
$$x = 10$$

SOLVE The number 7 is subtracted from x.
Add 7 to both sides to undo the subtraction.

CHECK To check your work, substitute 10 for x in the original equation and see if the equation is true. The solution is correct.

$$10 - 7 = 3$$
$$3 = 3 \text{ True}$$

Use opposite operations to "untangle" an equation. Remember, addition and subtraction are opposite operations, as are multiplication and division.

EXAMPLE 2   Solve: $6a = 5a - 4$

SOLVE Subtract $5a$ from both sides to move all the variables to the left side.

$$6a = 5a - 4$$
$$6a - 5a = 5a - 5a - 4$$
$$a = -4$$

CHECK
$$6(-4) = 5(-4) - 4$$
$$-24 = -20 - 4$$
$$-24 = -24 \text{ True}$$

In Example 2, positive $5a$ was moved from the right side of the equation to the left side. This was done by subtracting $5a$ from both sides. Another way to do this is to add $-5a$ to both sides. Remember, subtracting and adding the opposite are really the same thing.

Solve. Check your work.

1. $a + 4 = 7$ $b - 6 = 4$ $m - 2 = -5$

2. $n + 5 = -1$ $x + 3 = 2$ $y - 9 = -6$

3. $4a = 3a + 4$ $6m = 5m - 1$ $9x - 2 = 8x$

The following examples use multiplication and division to isolate the variables.

EXAMPLE 3   Solve: $9x = 63$

SOLVE   The number 9 is multiplied by x. Divide both sides of the equation by 9 to undo the multiplication.

$$9x = 63$$
$$\frac{9x}{9} = \frac{63}{9}$$
$$x = 7$$

CHECK $9(7) = 63$
$$63 = 63 \text{ True}$$

As you read the next example, remember that division in algebra is written as a fraction.

EXAMPLE 4   Solve: $\frac{a}{4} = 6$

SOLVE   The variable a is divided by 4. Multiply both sides of the equation by 4 to undo the division.

$$\frac{a}{4} = 6$$
$$4 \cdot \frac{a}{4} = 4 \cdot 6$$
$$a = 24$$

CHECK $\frac{24}{4} = 6$
$$6 = 6 \text{ True}$$

As you work with fractions and division, remember that dividing by a fraction is the same as multiplying by its reciprocal.

$$\frac{a}{b} \div \frac{c}{d} = \frac{a}{b} \times \frac{d}{c}$$

EXAMPLE 5   Solve: $\frac{4}{5}y = 20$

SOLVE   The variable y is multiplied by $\frac{4}{5}$. To divide by $\frac{4}{5}$, multiply by its reciprocal, $\frac{5}{4}$.

$$\frac{4}{5}y = 20$$
$$\frac{5}{4} \cdot \frac{4}{5}y = \frac{5}{4} \cdot 20$$
$$\frac{\cancel{20}}{\cancel{20}}y = \frac{5}{\cancel{4}} \cdot \frac{\cancel{20}^{5}}{1}$$
$$y = 25$$

CHECK $\frac{4}{5} \cdot 25 = 20$
$$\frac{4}{\cancel{5}_{1}} \cdot \frac{\cancel{25}^{5}}{1} = 20$$
$$20 = 20 \text{ True}$$

(For review of solving equations with fractions, see *Number Power Algebra,* page 66.)

Solve. Check your work.

4. $2a = -12$ $-6b = 42$ $-10m = -50$

5. $-5x = 25$ $-\frac{n}{3} = -5$ $\frac{y}{4} = -2$

6. $-\frac{5}{6}m = 25$ $\frac{2}{3}a = 12$ $-\frac{2}{5}x = -8$

Most algebra equations can be solved only after several steps. A strategy for solving equations is shown below. Remember, you won't need every step every time. Use common sense to decide which steps to use.

Plan for Solving Equations

STEP	WHAT TO DO
1. Simplify each side of the equation.	Eliminate grouping symbols, and group like terms.
2. Do any addition or subtraction steps.	Move the variables to one side of the equation and the number constants to the other side.
3. Do any multiplication or division steps.	Isolate the variable.
4. Check your solution.	Substitute your answer into the original equation to see if it makes the equation true.

Now put this strategy to work. Keep in mind that there is no one perfect way to solve an equation. The important thing is to use the rules of algebra systematically in ways that make sense in order to isolate the variable.

EXAMPLE 6   Solve: $20 + m = -2(m - 4)$

STEP 1 Simplify.
Multiply to remove the parentheses.

STEP 2 Add $2m$ to both sides to move the variables to the left side.
Then subtract 20 from both sides to move the constants to the right side.

STEP 3 Divide both sides by 3.
You have isolated the variable.

STEP 4 Replace the variable with your answer to check your work.

$$20 + m = -2(m - 4)$$
$$20 + m = -2m + 8$$
$$20 + m + 2m = -2m + 2m + 8$$
$$20 + 3m = 8$$
$$20 - 20 + 3m = 8 - 20$$
$$3m = -12$$
$$\frac{3m}{3} = \frac{-12}{3}$$
$$m = -4$$

$$20 + m = -2(m - 4)$$
$$20 + (-4) = -2(-4 - 4)$$
$$16 = -2(-8)$$
$$16 = 16 \text{ True}$$

Solve. Check your work.

7. $5x - 2 = 7(2x - 8)$ $18 - 3y = -5(y - 2)$

8. $11 + n = 5(n + 3)$ $-(1 - 8a) = -31 - 7a$

9. $5 - 3m = 5(1 + 7m)$ $2(b + 14) = -7(1 - b)$

10. $-4(2x - 5) - 8 = -12 - 5x$ $-2 + 7(a + 6) = -15 - 4a$

11. $2(m - 4) + m = -2(7 - 2m) - 6$ $-6 + 2(4x - 5) = 4(x + 2)$

12. $-4y - (5y - 3) = 4(4y - 5) - 2$ $-(b - 5) = -4(1 - b) - 6b$

So far, the equations you have solved each had exactly one solution. In the next two examples, you will see that some equations have no solution and other equations have infinitely many solutions.

EXAMPLE 7   Solve: $5x + 5 - 3x = 4 + 2x$

First combine like terms.
Then subtract $2x$ from both sides.
The variable is gone, and the remaining equation is never true!

$$5x + 5 - 3x = 4 + 2x$$
$$2x + 5 = 4 + 2x$$
$$2x + 5 - 2x = 4 + 2x - 2x$$
$$5 = 4 \text{ False}$$

This equation has **no solution.** The solution set for this equation is the **empty set**—a set that contains no members. You can show the empty set as { } or ∅. Both mean that no real number exists to make the solution true.

EXAMPLE 8   Solve: $4 - (2n + 6n) = -3n - 5n + 4$

Simplify and combine like terms.
The two sides are exactly the same.
If you add $8n$ to both sides, the variable disappears.
You are left with an equation that is always true.

$$4 - (2n + 6n) = -3n - 5n + 4$$
$$4 - 8n = 4 - 8n$$
$$4 - 8n + 8n = 4 - 8n + 8n$$
$$4 = 4 \text{ True}$$

This equation has infinitely many solutions. Any real number that you substitute for n will make the equation true. This kind of equation is called an **identity.** The solution set is the set of **all real numbers.**

Don't be confused by equations that have zero as a solution. The number 0 is a real number. When a variable equals zero, the variable has a value, and its value is 0.

EXAMPLE 9   Solve: $n - 3 = -3$

Add 3 to both sides.
The equation has one solution—the number 0.

$$n - 3 = -3$$
$$n - 3 + 3 = -3 + 3$$
$$n = 0$$

Solve.

13. $-4(1 + 3b) = -4$ $\qquad$ $n + 3 - 4n = 3 - 2n - n$ $\qquad$ $a - 6 + 3a = -5 + 4a$

14. $2(y + 3) = -3y + 5y$ $\qquad$ $-(d - 4) = 4$ $\qquad$ $3 = -(6x - 3) + 6x$

15. $-2b = 4b - 6b$ $\qquad$ $-2 - 4a + 5a = -8 + a$ $\qquad$ $7y - 3 = 4y - 3 + 3y$

Equations containing fractions or decimals may seem more difficult. Study the next two examples to see how to clear fractions and decimals from equations.

EXAMPLE 10 Solve: $a - \dfrac{5}{3} = \dfrac{10}{9} + \dfrac{1}{6}a$

STEP 1 Clear the fractions by multiplying both sides of the equation by the lowest common denominator of the fractions. In this case, the LCD is 18.

$$18\left(a - \frac{5}{3}\right) = 18\left(\frac{10}{9} + \frac{1}{6}a\right)$$

$$18a - \frac{\overset{6}{\cancel{18}}}{1} \cdot \frac{5}{\cancel{3}} = \frac{\overset{2}{\cancel{18}}}{1} \cdot \frac{10}{\cancel{9}} + \frac{\overset{3}{\cancel{18}}}{1} \cdot \frac{1}{\cancel{6}}a$$

$$18a - 30 = 20 + 3a$$

$$15a = 50$$

STEP 2 Solve by isolating the variable. Subtract $3a$ from both sides; then add 30 to both sides. Divide both sides by 15.

$$a = \frac{50}{15}$$

$$a = \frac{10}{3}$$

STEP 3 Write the fraction in lowest terms.

Note: In algebra, improper fractions are used instead of mixed numbers. The answer $a = \frac{10}{3}$ is acceptable. You do not need to write it as $3\frac{1}{3}$ unless it makes sense in the context of the problem.

To clear decimals from an equation, you can multiply both sides of the equation by a power of ten. Look for the number that has the greatest number of decimal places. Then multiply the equation by that power of 10.

EXAMPLE 11 Solve: $4.4m + 1.4 = -3.64 - 1.2m$

STEP 1 Clear the decimals. -3.64 has 2 decimal places, so multiply by the 2nd power of 10. $10^2 = 100$, so multiply each side by 100.

$$100(4.4m + 1.4) = 100(-3.64 - 1.2m)$$
$$440m + 140 = -364 - 120m$$
$$560m = -504$$

STEP 2 Solve by isolating the variable. Then simplify.

$$m = \frac{-504}{560}$$

$$m = -0.9$$

Do you have to clear the fractions and decimals to solve an equation? No. You can isolate the variable by performing opposite operations with fractions and decimals. Many students, however, find they make fewer errors if they clear fractions and decimals first.

Solve for the variable.

16. $\frac{5}{3}n + 1 = \frac{4}{3}n + \frac{16}{15}$ $\frac{3}{2}x - x = -\frac{5}{3} - 2x$

17. $y - 0.9 = -0.5y$ $2.06 + 5.5m = 2.4 + 2.1m$

18. $-\frac{20}{9} + \frac{2}{3}x - 3x = -\frac{3}{2}x$ $\frac{1}{4}a + \frac{7}{5} = \frac{9}{10} - a + \frac{3}{4}a$

19. $-12.7 - 2.4b = 0.3b + 0.8$ $1 + 5.7n = -1.34 + 1.8n$

20. $\frac{1}{2} = \frac{5}{4}\left(-\frac{2}{5}x + 1\right)$ $-1.5(1.8y + 6) = -6.03$

(**Hint:** Use the distributive property to remove the parentheses before you clear the fractions or decimals.)

   Synthesis   Some equations that look complicated can be made simpler with a little manipulation and knowledge of equation rules. Try substituting a variable for a difficult part of the equation.

EXAMPLE   Solve: $2\sqrt{x} - 3 = 12 - \sqrt{x}$

You haven't learned how to deal with radicals in equations yet, but see if you can apply some common sense to find the solution. The first step is to choose a variable to substitute for $\sqrt{x}$ in the equation, so let $\sqrt{x} = y$.

Replace $\sqrt{x}$ with y. The equation is now simple to solve. $2y - 3 = 12 - y$

Now return to the first step. If $\sqrt{x} = y$ and $y = 5$, then $\sqrt{x} = 5$ (by the transitive rule). $3y = 15$

$y = 5$

Think About It: The square root of what number is 5? The variable x must equal **25**.

Solve these equations by replacing the radicals with variables.

21. $2\sqrt{m} + 6 - \sqrt{m} = 22 - 3\sqrt{m}$ $5 - \sqrt{3x} = 3\sqrt{3x} - 7$

Solving Word Problems

A word problem is a chance to apply the algebra you have learned to a real situation. An algebra word problem sounds a little like a puzzle. There are clues about the numbers in the situation. To solve the puzzle, find a way to write the clues as algebraic expressions and equations, and then solve for the missing number. To make the process easier, follow the steps in the plan below.

5-Step Plan for Solving Word Problems

STEP 1 Read and organize.	Carefully read the information in the problem. You may want to create a chart or list showing every quantity in the problem.
STEP 2 Write expressions.	Decide which quantity in the chart should be x. Then write expressions for all the parts of the problem using x.
STEP 3 Write an equation.	Combine the expressions to create an equation. Make sure the equation represents the situation described.
STEP 4 Solve and check.	Solve the equation for x; then check your work.
STEP 5 Apply the answer.	Go back to the original question. The answer may be the value of x, or you may need to use the value of x to answer the question.

The following problem describes number relationships, a common application of algebra. Notice how the 5-step plan is used to answer the question.

EXAMPLE 1 In a game, Leigh scored 50 points less than twice the points Fran scored. Together they scored 220 points. How many points did Leigh score?

STEP 1 Read carefully. There are three quantities—two numbers and their sum. Make a chart with those headings.

Leigh's points	Fran's points	Sum of their points
$2x - 50$	x	220

STEP 2 Let x represent Fran's points. Leigh's points are 50 points less than twice the points Fran scored, or $2x - 50$. The sum is given: 220 points.

STEP 3 Write an equation. Combine the expressions representing the points and set it equal to the sum.

$$2x - 50 + x = 220$$
$$3x - 50 = 220$$
$$3x = 270$$
$$x = 90$$

Check:
$$2(90) - 50 + 90 = 220$$
$$180 - 50 + 90 = 220$$
$$130 + 90 = 220$$
$$220 = 220$$

STEP 4 Solve for x and check.

STEP 5 Return to the original problem. You need to know Leigh's score, not Fran's score. Use the value of x in the expression that represents Leigh's points.

$$2x - 50 = 2(90) - 50$$
$$= 180 - 50$$
$$= \textbf{130 points}$$

The previous example shows how useful a chart can be. Without the chart, you might have answered 90, the value of x. In a multiple-choice problem, 90 would probably be a choice. Creating a chart makes it clear that $x = 90$ isn't the final step, and it tells exactly what to do to find the answer.

Solve each word problem. For problems 1 and 2, charts have been started for you.

1. Mark's age is 7 years less than twice Maria's age. The sum of their ages is 50. How old is Mark?

Mark's age	Maria's age	Sum of their ages
	x	50

2. The sum of three consecutive odd numbers is 231. What is the smallest number?

1st	2nd	3rd	Sum
x	$x + 2$		231

(**Hint:** If the first odd number is x, the second odd number is $x + 2$.)

3. One number is 5 less than another number. If the sum of the first number and twice the second is 97, what are the two numbers?

4. When one-third of a number is decreased by 12, the result is 23. What is the number?

5. Find two consecutive odd integers such that the first is three more than twice the second.

(**Hint:** The word *integers* includes negative numbers.)

6. The sum of four consecutive even numbers is 196. What is the largest of the four numbers?

(**Hint:** The value of x may not be the answer to the question.)

7. Stuart is twice as old as his brother Brad. Their older sister Cathy is 2 years younger than twice Stuart's age. If the sum of all their ages is 33, how old is Stuart?

8. Four times an integer is equal to 6 times the next consecutive integer. What are the two integers?

9. At the end of a game, Elio had scored 3 times as many points as Erika. Zach scored 6 points fewer than double the number Erika scored. Together Elio and Erika scored 60 points. How many points did Zach score?

10. Lina scored a 76 on the first of three tests. Her second test score was 1 point lower than her third test score. If her average for the three tests was 87, what did she score on the third test?

Solving Geometry Problems

Algebra problems are sometimes based on geometry. To solve these problems, you need to combine your understanding of geometry with your ability to write and solve equations. Review these facts from geometry.

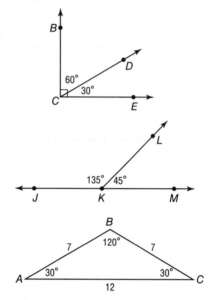

A **right angle** measures 90°. ∠BCE is a right angle.

Two angles are **complementary** if their sum is 90°. ∠BCD and ∠DCE are complementary.

A **straight angle** looks like a line and measures 180°. ∠JKM is a straight angle.

Two angles are **supplementary** if their sum is 180°. ∠JKL and ∠LKM are supplementary.

The sum of the interior angles of any triangle is 180°.

An **isosceles triangle** has two equal sides and two equal angles. △ABC is an isosceles triangle.

Now use these geometry facts to solve an algebra problem.

EXAMPLE 2 In △CDE, ∠C is three times larger than ∠D. ∠E is 12° less than four times ∠D. What is the measure of ∠C?

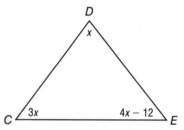

STEP 1 Read carefully. This problem is about a triangle, but you aren't given a diagram. Go ahead and draw one.

The drawing is not to scale, and don't worry if yours is not either. The drawing is just to keep the facts of the problem organized.

STEP 2 ∠C is described in terms of ∠D. ∠E is described in terms of ∠D. The angle that is not described is ∠D, so let the variable x represent the measure of this angle. Now write expressions for the remaining angles: $m\angle C = 3x$ and $m\angle E = 4x - 12$.

STEP 3 Write an equation and solve for x.

$$x + 3x + 4x - 12 = 180$$
$$8x - 12 = 180$$
$$8x = 192$$
$$x = 24$$

STEP 4 Now use the value of x to find the measures of the other angles. To check your answer, make sure the sum of all the angles is 180°.

$$m\angle D = x$$
$$= 24$$

$$m\angle C = 3x$$
$$= 3(24)$$
$$= 72$$

STEP 5 The problem asks for the measure of ∠C. $m\angle C = \mathbf{72°}$

$$m\angle E = 4x - 12$$
$$= 4(24) - 12$$
$$= 96 - 12$$
$$= 84$$

Check: 72
    24
  +84
  ———
   180

Solve the problems. For problems 1 and 2, diagrams are already drawn.

1. In a triangle, one angle is 3 times the measure of a second angle. A third angle is 15° smaller than the second angle. What is the measure of the largest angle?

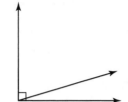

2. ∠A and ∠B are complementary. ∠A measures 10° greater than four times the measure of ∠B. What is the measure of ∠B?

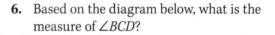

3. Triangle *DEF* is an isosceles triangle. The measure of ∠E is 12° more than the sum of angles *D* and *F*. If m∠D = m∠F, what is the measure of ∠E?

   (**Hint:** m∠D is read *the measure of angle D.*)

4. Angles *J* and *K* are supplementary angles. ∠J is 21° less than twice the measure of ∠K. What is the measure of ∠J?

5. The equal legs of an isosceles triangle are each 1 cm longer than twice the length of the base. If the perimeter of the triangle is 22 cm, what are the lengths of the three sides of the triangle?

   (**Hint:** Perimeter is the distance around a figure.)

6. Based on the diagram below, what is the measure of ∠BCD?

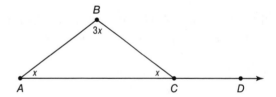

7. One side of a triangle is 5 inches longer than a second side. A third side is 8 inches shorter than twice the second side. If the perimeter of the triangle is 65 inches, what is the length of the shortest side?

8. **Challenge** The length of a rectangle is 5 cm longer than its width. If both the length and width are increased by 3 cm, the area is increased by 48 sq cm. What are the dimensions of the original rectangle?

Solving More Word Problems

Many algebra problems are about everyday things such as shopping, driving, and earning a living. Remember, the principles for algebraic problem solving are the same no matter the situation. Find a way to represent parts of the problem with variables and symbols. Then combine the parts to create and solve an equation.

EXAMPLE 1   A tip jar contains 60 coins, all either dimes or nickels. The total value of the coins is $4.15. How many of the coins are nickels?

STEP 1   Consider two things: the value of the coins and the number of coins. Your chart needs two rows.

	Dimes	Nickels
Number	x	$60 - x$
Value	$10x$	$5(60 - x)$

STEP 2   If x is the number of dimes, then $60 - x$ is the number of nickels. Now express each value by multiplying the worth of the coin by the number of coins.

It is simpler to express the values in cents rather than dollars: $10x$ instead of $0.1x$.

STEP 3   Write an equation. Make sure to express the total in cents as well: 415 cents instead of $4.15.

$$10x + 5(60 - x) = 415$$

STEP 4   Solve for x. Look at the chart to see that x is the number of dimes.

$$10x + 300 - 5x = 415$$
$$5x = 115$$
$$x = 23$$

STEP 5   Return to the original problem. The problem asks for the number of nickels, which is represented by $60 - x$.

$$60 - 23 = \textbf{37 nickels}$$

Step 6 would be to check your answer—you can do that on your own.

The next example is a typical puzzle problem. The challenge is how to represent ages in the present and the past using only one variable.

EXAMPLE 2   Samantha is four years older than her brother Dan. Five years ago, she was twice as old as her brother. How old are they now?

STEP 1   Consider their ages now and 5 years ago.

	Now	5 years ago
Samantha	x	$x - 5$
Dan	$x - 4$	$x - 4 - 5$

STEP 2   If x is Samantha's age now, $x - 4$ is Dan's age now. Use these facts to write expressions for their ages 5 years ago.

STEP 3   Write an equation using the information that 5 years ago, Samantha's age was twice her brother's age.

$$x - 5 = 2(x - 4 - 5)$$
$$x - 5 = 2(x - 9)$$

STEP 4   Solve for x.

$$x - 5 = 2x - 18$$
$$-x = -13$$
$$x = 13$$

STEP 5   Return to the original problem, and answer the question. **Samantha is 13 now. Dan is 9** (13 − 4) now.

Check that your answer makes sense. Five years ago, Samantha would have been 8 and Dan would have been 4. Since 8 is twice 4, the answer makes sense.

Solve each word problem. For problems 1 and 2, charts have been started for you.

1. Tickets for a school play are $12 for adults and $8 for students. The school sold 200 tickets for Friday night. If the school earned $1,860 for the show, how many $12 tickets did they sell?

	Adult	Student
Number	x	200 − x
Price		

2. Abe is 8 years older than Ben. Four years ago, Abe was twice Ben's age. How old is Ben now?

	Now	4 years ago
Abe		
Ben		

3. Max is three years younger than twice what his age was six years ago. How old is Max now?

4. A cashier has 90 coins, all quarters and dimes. If the total value of the coins is $14.40, how many quarters does the cashier have?

5. Leon earns $8 an hour at a part-time job during the week and $12 per hour working on the weekends. Last week Leon worked 28 hours and earned $276. How many hours did he work on the weekend?

6. Alarie and Ryan went on a 3-day trip. The second day they drove 50 miles more than they drove on the first day. The third day they drove $\frac{1}{3}$ the total distance of the first two days. If they drove 920 miles in all, what distance did they drive on the third day?

7. On a test, some questions were worth 8 points each and some were worth 3 points each. Omar answered 23 questions right and had a score of 94. How many 3-point questions did Omar answer correctly?

8. Three basketball players scored a total of 45 points. The center scored 5 points more than the guard. The forward scored 5 points fewer than three times the guard's points. How many points did the forward score?

9. Roshni has 30 coins in her backpack, all nickels, dimes, and quarters. She has a total of $4.30. If Roshni has 5 more dimes than nickels, how many quarters does she have?

10. **Challenge** The numerator of a fraction is 4 less than its denominator. If you add 1 to both the numerator and the denominator, the resulting fraction is equal to $\frac{2}{3}$. What is the original fraction?

Inequalities

In this lesson, you will learn to

- Solve inequalities with one variable.
- Solve compound inequalities.

Most of the rules of equality can be applied to inequalities. The goal in solving an inequality is to isolate the variable by removing grouping symbols, combining like terms, and using opposite operations.

In this next example, notice how the steps in solving the inequality are the same steps you would use to solve an equation.

EXAMPLE 1   Solve $5x + 3 \leq -12$ and solve $5x + 3 = -12$.

Use opposite operations.

STEP 1   Add -3 to both sides.

STEP 2   Then divide both sides by 5.

$$5x + 3 \leq -12 \qquad\qquad 5x + 3 = -12$$
$$5x \leq -15 \qquad\qquad\quad 5x = -15$$
$$\frac{5x}{5} \leq \frac{-15}{5} \qquad\qquad\quad \frac{5x}{5} = \frac{-15}{5}$$
$$x \leq -3 \qquad\qquad\qquad x = -3$$

Some rules of equality, however, do not apply to inequalities. For instance, the symmetric rule of equality says that if $a = b$, then $b = a$. Would that work with an inequality? You know that $-3 < 0$. Can you switch the sides and say $0 < -3$? No, the symmetric rule of equality doesn't apply. If you switch the sides of an inequality, you need to also reverse the symbol. Instead the rule is: If $a < b$, then $b > a$.

You must reverse the inequality symbols when you are

- Multiplying or dividing both sides by a negative number.
- Exchanging the sides of an inequality.

> **Think About It:** Why does multiplying or dividing by a negative number reverse the sign?
>
> $3 < 5$ is a true fact. Multiply both sides by -1. Is the result true?

EXAMPLE 2   Solve: $-2(a + 5) > 2$

STEP 1   Use the distributive property to remove the parentheses.

$$-2(a + 5) > 2$$
$$-2a - 10 > 2$$

STEP 2   Add 10 to both sides.

$$-2a > 12$$

STEP 3   Divide both sides by -2 AND reverse the sign.

$$\frac{-2a}{-2} < \frac{12}{-2}$$
$$a < -6$$

How do you check an inequality? Look at the last example. Because a is less than -6, replace a in the original inequality with any number less than -6. The number -7 is used in the check at the right. The answer checks because the inequality is true.

$$-2(a + 5) > 2$$
$$-2(-7 + 5) > 2$$
$$-2(-2) > 2$$
$$4 > 2 \text{ True}$$

There are actually many solutions for the inequality $-2(a + 5) > 2$. This check only proves that -7 is a solution. Other solutions include -8, -20, and -100. Since there are infinitely many solutions, the solution set is often shown as a graph on a number line.

To graph the solution set of $-2(a + 5) > 2$, graph the final statement: $a < -6$.
Remember, the open circle means that -6 is not a solution.

 Math Talk: $5 > x$ can be read *5 is greater than x* or it can be read *x is less than 5*.

Solve and graph each inequality. Check your work.

1. $3n - 5 < -9 + n$

$-10 \geq -2(1 - 4x)$

$7 + 3a < 6a + 19$

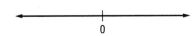

2. $7 \leq 3y - (y - 1)$

$-2(3 - m) < -2$

$6(1 + 4n) < n + 6$

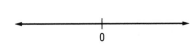

3. $26 \geq 2(4b + 5)$

$x - 1 - 4x < 3x - 1$

$-7y + 3(y - 5) \leq 17 - 8y$

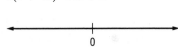

(**Hint:** To work with fractions, first remove any parentheses using the distributive property; then clear the fractions by multiplying both sides by the LCD.)

4. $-\dfrac{5}{4}x - x > 0$

$7 \leq -\dfrac{5}{2}\left(\dfrac{12}{5}a + 2\right)$

$6y + 7 > -\dfrac{4}{3}\left(2y + \dfrac{5}{4}\right)$

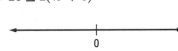

In real life, two inequalities are often combined in one idea. Suppose you want to work at least 15 hours per week at a part-time job, but not more than 20 hours per week. You could express your desired hours (h) as $15 \leq h \leq 20$. This is an example of a **compound inequality.** The variable h is greater than or equal to 15 AND less than or equal to 20.

To graph $15 \leq h \leq 20$ on a number line, think: h is between 15 and 20, so shade the part of the line between 15 and 20. Use closed circles because the graph includes 15 and 20.

EXAMPLE 3   Solve and graph: $2 < x + 3 < 6$

STEP 1   Isolate the variable between the outer expressions by subtracting 3 from all three parts of the inequality.

$$2 < x + 3 < 6$$
$$2 - 3 < x + 3 - 3 < 6 - 3$$
$$-1 < x < 3$$

STEP 2   Graph the solution.

This kind of compound inequality is called a **conjunction** because the two ideas are joined by the word *and*.

Both circles are open because the values −1 and 3 are not included in the solution set.

In Example 3, the two ideas are $x + 3 > 2$ and $x + 3 < 6$. As shown at the right, you could solve each inequality separately and arrive at the same answer.

$$x + 3 > 2 \qquad x + 3 < 6$$
$$x > -1 \qquad x < 3$$

When you combine the results, $x > -1$ AND $x < 3$ becomes $-1 < x < 3$. Notice that you need to exchange the sides of the first inequality to combine the two ideas.

Another kind of compound inequality is called a **disjunction** because the two ideas are joined by the word *or*. In a disjunction only one part of the inequality is true at a time, while in a conjunction, both parts are true at the same time.

EXAMPLE 4   Solve and graph: $x + 2 \leq 0$ or $x - 5 > 3$

STEP 1   Solve both parts separately.

STEP 2   Write both inequalities separated by *or*. The answer is $x \leq -2$ or $x > 8$.

$$x + 2 \leq 0 \qquad x - 5 > 3$$
$$x \leq -2 \qquad x > 8$$

STEP 3   Graph the solution.

Interval notation is a helpful system for recording the solution set of an inequality.

Consider the graph of the inequality $-4 \leq x < 3$.

What details do you need to understand the graph? You need to know the beginning and end points of the graph and whether the circles are open or closed.

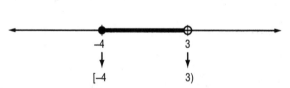

A bracket represents a closed circle.

A parenthesis represents an open circle.

Those facts can be represented with interval notation. A bracket symbol means the point is included in the set. A parenthesis means it is not included in the set.

The interval notation for the graph is **[−4, 3)**.

Not all graphs have endpoints. The graph of the inequality $x \geq 2$ is a closed circle at 2 and an arrow pointing to the right. This shows that the solution set continues to positive infinity. In interval notation the infinity symbol, ∞, is used; $x \geq 2$ is written $[2, \infty)$.

EXAMPLE 5 Express the inequality $x < 0$ using interval notation.

STEP 1 Draw the graph.

STEP 2 Write symbols to represent the graph. Use the
symbol $-\infty$ to represent the infinity that extends
to the left side of the number line. The inequality
is written **$(-\infty, 0)$**.

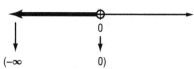

Always use a parenthesis next to
∞ and $-\infty$.

A disjunction requires an additional step. Both parts of the graph must be
represented separately and then joined with the union symbol ∪.

EXAMPLE 6 Express $x < 1$ or $x \geq 4$ using interval notation.

STEP 1 Draw the graph.

STEP 2 Express both parts of the graph separately. Then join
the two parts with the union symbol:
$(-\infty, 1) \cup [4, \infty)$

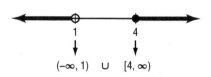

Solve and graph each inequality. Then write your answer in interval notation.

5. $-12 < -4x \leq 0$

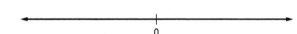

$6 + n \geq 7$ or $n + 2 \leq -5$

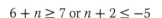

6. $-13 \leq b - 8 < -11$

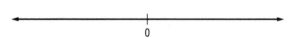

$-2 \leq \frac{x}{4} < 0$

7. $-3m < -9$ or $\frac{m}{9} < -1$

$-4 \leq 5 + x < 7$

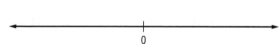

8. $8 + 5b \geq 38$ or $-7 + 4b < 13$

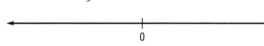

$8 < -1 + \frac{9}{2}x < \frac{25}{2}$

9. $-a - 2 \leq -9$ or $4 + 5a < 9$

$5 + 9y \leq -31$ or $2y - 7 > -7$

10. $-13 \leq 6x - 1 < 5$

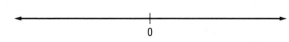

$9 < 6m + 3 \leq 21$

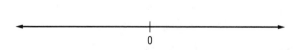

Absolute Value Equations

In this lesson, you will learn to

- Evaluate absolute value expressions.
- Solve absolute value equations and inequalities.

The absolute value of a number is its distance from 0 on a number line. Putting this to practical use, imagine you start at your home and drive 4 miles east. How far are you from home? You are 4 miles away. Now imagine that you start at home and drive 4 miles west instead. How far are you from home? Are you −4 miles from home? No, you are still 4 miles away.

The absolute value of −4 is 4.

The absolute value of a number is either a positive number or 0. The symbol $|n|$ represents "the absolute value of n."

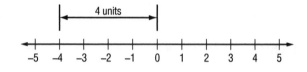

EXAMPLES $|2| = 2$ $|-2| = 2$ $|0| = 0$ $-|2| = -2$

Absolute value bars act as grouping symbols when placed around expressions. In the order of operations, treat absolute value bars as you would parentheses.

EXAMPLE 1   Simplify: $|3 - 6| + |-7 + 2|$

STEP 1 Do the operations within the absolute value bars.

STEP 2 Find the absolute values.

STEP 3 Complete any operations.

$$|3 - 6| + |-7 + 2| = |-3| + |-5|$$
$$= 3 + 5$$
$$= 8$$

Simplify.

1. $3|-6| - 10$ $7 - |4 - 10|$ $-|3 - 8|$

2. $|-12| + |8 - 2|$ $5|1 - 3|$ $-|9 - 7| \cdot -3|-2|$

3. $-10 + |2 - 12|$ $6 + |-5 - 5|$ $-|-7 + 5| \cdot |-3|$

Equations sometimes contain absolute value expressions. Think about the equation: $|x| = 4$. What are the possible values of x? You know that absolute value is the distance from 0. On the number line on this page you can see that there are two values of x that are 4 units from 0. The possible solutions are 4 and −4. Look at how this works in an equation with more than one step.

EXAMPLE 2   Solve: $|a - 5| = 3$

STEP 1 Remove the absolute value symbols and write two equations. Make the right side of the equation positive in one and negative in the other.

$$a - 5 = 3 \qquad a - 5 = -3$$
$$a = 8 \qquad\quad a = 2$$

STEP 2 Solve both equations. The two solutions are **2** and **8**.

When an equation contains an operation outside the absolute value symbol, first isolate the absolute value expression on one side of the equation. Then solve.

EXAMPLE 3   Solve: $|1 - 4m| + 7 = 14$

STEP 1   Subtract 7 from both sides to isolate the absolute value expression.

$$|1 - 4m| + 7 = 14$$
$$|1 - 4m| = 7$$

STEP 2   Remove the absolute value symbols and write two equations.

$$
\begin{array}{ll}
1 - 4m = 7 & 1 - 4m = -7 \\
-4m = 6 & -4m = -8 \\
m = -\frac{6}{4} & m = \frac{-8}{-4}
\end{array}
$$

STEP 3   Solve both equations.

$$
\begin{array}{ll}
m = -\frac{3}{2} & m = 2
\end{array}
$$

The two solutions are $-\frac{3}{2}$ and **2.**

Check your answer by substituting each solution for m in the original absolute value equation. Make sure each solution makes the equation true. Some absolute value equations will have no solutions, and some will have only one. Study the two examples below to see how to recognize these special situations.

EXAMPLE 4   Solve: $|m + 6| + 2 = 1$

STEP 1   Subtract 2 from both sides to isolate the absolute value expression.

$$|m + 6| + 2 = 1$$
$$|m + 6| = -1$$

STEP 2   Examine the equation. How could an absolute value expression ever equal a negative number? It couldn't. This equation has **no solution.**

$|m + 6| = -1$ has **no solution.**

EXAMPLE 5   Solve: $|x + 4| - 3 = -3$

STEP 1   Add 3 to both sides.

STEP 2   Remove the absolute value symbols and write two equations. Wait! Because 0 and −0 are really the same value, you can write only one equation. There will be only one solution, **−4.**

$$|x + 4| - 3 = -3$$
$$|x + 4| = 0$$
$$x + 4 = 0$$
$$x = -4$$

Solve.

4. $|-7 + 5b| = 8$ $|4n - 9| = 9$ $|-2m + 6| = -12$

5. $|x - 4| + 5 = 8$ $|2a - 9| + 4 = 21$ $-2 - |10 - 5y| = -7$
                                                                   (**Hint:** Multiply every term by −1.)

6. $|3c + 6| = 0$ $-10|n - 6| = -40$ $2 + |2x + 2| = 16$

Absolute Value Inequalities

What happens when an absolute value expression is part of an inequality? For example, what are the solutions for $|x| < 5$? The solution set must be all the numbers whose distance from 0 is less than 5. In other words, the solutions must be all the numbers between −5 and 5. The opposite is true for the inequality $|x| > 5$. Its solution set is all the numbers less than −5 and greater than 5.

Learn the following patterns:

Less than pattern: If $|a| < b$, then $-b < a < b$.

Greater than pattern: If $|a| > b$, then $a < -b$ or $a > b$.

EXAMPLE 6   Solve and graph: $|x - 6| \leq 3$

STEP 1   Remove the absolute value symbol, and solve with the right side of the inequality as positive.

$$x - 6 \leq 3$$
$$x \leq 9$$

STEP 2   Write a second inequality. Make the right side of the inequality negative AND reverse the symbol. Solve.

$$x - 6 \geq -3$$
$$x \geq 3$$

STEP 3   The original inequality symbol is *less than or equal to* ($\leq$), so the solution set is between the endpoints. Graph.

The solution set for x is $\mathbf{3 \leq x \leq 9}$, or written in interval notation: $\mathbf{[3, 9]}$.

Before deciding if an inequality is a *less than* or *greater than* inequality, first isolate the absolute value expression on the left side of the inequality.

EXAMPLE 7   Solve and graph: $2 - 7|m - 2| < -26$

STEP 1   Isolate the absolute value expression by subtracting 2 from both sides and then dividing both sides by −7. **Remember:** when you divide by a negative number, you must reverse the inequality symbol.

$$2 - 7|m - 2| < -26$$
$$-7|m - 2| < -28$$
$$|m - 2| > 4$$

STEP 2   Examine the inequality. The inequality has a *greater than* symbol, so the solution set lies beyond the endpoints.

$$m - 2 > 4 \qquad m - 2 < -4$$
$$m > 6 \qquad\qquad m < -2$$

STEP 3   Write and solve two inequalities. Remember to reverse the inequality symbol when you change the sign of the right side.

STEP 4   The solution set is a disjunction. Graph.

The solution set for m is $\mathbf{m < -2}$ or $\mathbf{m > 6}$, also written: $\mathbf{(-\infty, -2) \cup (6, \infty)}$.

Watch out for special situations. An absolute value inequality may have no solution, or the solution set may be the set of all real numbers. For the next two examples, recall that the absolute value of a number is always a positive number or 0.

EXAMPLE 8 Solve: $|y| < -2$

Could the distance from 0 ever be less than a negative number? No. This inequality has **no solutions.**

The solution set is **Ø.**

The graph is a blank number line.

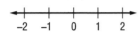

EXAMPLE 9 Solve: $|n| > -4$

Could the distance from 0 ever be greater than a negative number? Yes. The absolute value of any number must <u>always</u> be greater than a negative number.

The solution set is **all real numbers: $(-\infty, \infty)$.**

The graph includes all values.

Solve and graph each inequality. Then write your answer in interval notation.

7. $|3n| \leq 6$

0

$|a - 5| > 5$

0

8. $\left|\dfrac{x}{2}\right| > 4$

0

$|b + 2| < 2$

0

9. $|y - 4| + 1 < 0$

0

$-4\left|\dfrac{m}{2}\right| - 6 > -8$

0

10. $\dfrac{|x + 5|}{3} \leq 1$

0

$8 + 4|-7a| \geq -48$

0

11. $5 + 2|-2 - y| > 11$

0

$\left|\dfrac{b}{4}\right| - 1 > -3$

0

12. $4 - |3x - 3| > -5$

0

$|4y - 8| + 5 > 9$

0

Formulas: Solving for a Specific Variable

In this lesson, you will learn to

- Solve literal equations for a specific variable.

A **literal equation** involves more than one variable. Most literal equations are formulas that show the process for finding a particular quantity. For example, to find the area of a rectangle, you multiply length and width. That process is recorded as $A = lw$. The formula makes it easy to remember the process for finding area.

Sometimes you know the area and the length of a rectangle, and you need to find the width. Of course, you can simply plug in the numbers you know and solve for the missing one. Another approach is to rewrite the formula so that it solves for the measurement you need. This approach is very useful when entering information in a graphing calculator or a computer spreadsheet.

Many formulas are simply a product of several variables. These formulas can be easily rewritten and solved for the missing variable.

EXAMPLE 1 The formula used to solve for the lateral surface area (S) of an open cylinder is $S = 2\pi rh$, where r represents radius and h represents height. Solve the formula for r.

STEP 1 Isolate r by dividing both sides of the equation by $2\pi h$. Then cancel.

STEP 2 Exchange the sides of the equation.

$$S = 2\pi rh$$
$$\frac{S}{2\pi h} = \frac{2\pi rh}{2\pi h}$$
$$r = \frac{S}{2\pi h}$$

Canceling works because $\frac{2\pi h}{2\pi h} = 1$.

Formulas with addition and subtraction steps require more work.

EXAMPLE 2 The formula used to write the slope-intercept form for a line is $y = mx + b$. Solve the formula for x.

STEP 1 To isolate x, first perform any addition or subtraction. Subtract b from both sides.

STEP 2 Divide both sides of the equation by m, and cancel.

STEP 3 Exchange the sides of the equation.

$$y = mx + b$$
$$y - b = mx + b - b$$
$$y - b = mx$$
$$\frac{y - b}{m} = \frac{mx}{m}$$
$$x = \frac{y - b}{m}$$

More complicated formulas require multiple steps. The following plan can help you decide how to approach a literal equation.

1. Clear fractions using multiplication.
2. Use the distributive property to remove grouping symbols.
3. Perform any addition or subtraction.
4. Divide both sides of the equation by the same numbers or variables to make canceling possible and to isolate the variable.
5. Exchange the sides of the equation if necessary.

EXAMPLE 3   Solve the formula $D = \frac{C - S}{n}$ for S.

STEP 1   Multiply both sides of the formula by n to clear the fraction. Cancel.

STEP 2   Subtract C from both sides.

STEP 3   Multiply both sides of the formula by -1.

STEP 4   Exchange the sides.

$$D = \frac{C - S}{n}$$
$$Dn = \frac{n(C - S)}{n}$$
$$Dn = C - S$$
$$Dn - C = -S$$
$$C - Dn = S$$
$$S = C - Dn$$

Take another look at Example 3. Did you notice that these were the same steps you would take to solve for S if the values of the variables were known?

EXAMPLE 4   Let $C = 50$, $n = 5$, and $D = 8$. Using the original formula, how would you solve for S?

STEP 1   Substitute the values into the formula.

STEP 2   Multiply both sides by 5.

STEP 3   Subtract 50 from both sides.

STEP 4   Multiply both sides of the equation by -1.

STEP 5   Exchange the sides.

$$D = \frac{C - S}{n}$$
$$8 = \frac{50 - S}{5}$$
$$40 = 50 - S$$
$$-10 = -S$$
$$10 = S$$
$$S = 10$$

The steps are the same. If you have trouble solving a literal equation, ask yourself: How would I solve this problem if there were numbers in place of the variables?

Solve for the indicated variable.

1.   Solve $P = 2l + 2w$ for w.   Solve $ax + by = c$ for y.

2.   Solve $ax + b = 0$ for x.   Solve $V = \frac{1}{3}bh$ for b.

3.   Solve $A = P(1 + rt)$ for r.   Solve $A = \frac{1}{2}(b_1 + b_2)h$ for b_1.

4.   Solve $F = \frac{9}{5}C + 32$ for C.   Solve $D = \frac{R(100 - x)}{100}$ for R.

5. Solve $A = T(S - R)$ for R.

Solve $V = \frac{1}{3}\pi r^2 h$ for h.

6. Solve $C = \frac{pV}{T}$ for p.

Solve $C = \frac{5}{9}(F - 32)$ for F.

7. Solve $D = 180(n - 2)$ for n.

Solve $\frac{t - h}{a} = I$ for h.

Synthesis   When using a graphing calculator to graph linear equations, you must be able to write equations in terms of y. The phrase *in terms of y* means that the equation must begin with the expression $y =$.

EXAMPLE 5   Rewrite $2x + 3y = 7$ in terms of y.

STEP 1   Subtract $2x$ from both sides of the equation.

STEP 2   Divide both sides by 3.

STEP 3   Write the equation in standard form. Notice the term with x is written first, followed by the constant term.

$$2x + 3y = 7$$
$$3y = 7 - 2x$$
$$y = \frac{7}{3} - \frac{2x}{3}$$
$$y = -\frac{2}{3}x + \frac{7}{3}$$

Write each equation in terms of y.

8. $5y + 2x = 20$

$4x + 8y = 3$

9. $\frac{x}{2} = \frac{y}{3}$

$2y - \frac{1}{2}x = 8$

10. $5x - y = -2$

$3y - x^2 = 9x - x^2$

11. $\frac{2x - y}{2} = \frac{1 - y}{3}$

Challenge $y^2 - x^2 = -2$

(**Hint:** Take the square root of both sides to undo a power of 2.)

Equation and Inequality Basics Review

Solve the problems below. When you finish, check your answers at the back of the book and correct any errors.

Graph the solution set for each equation or inequality. Add points and markings to the graph as needed.

1. $x = -3$

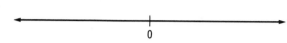

$y = -2$ and $y = 2$

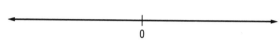

2. $a < -4$

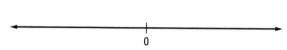

$b \geq -1$

3. $-5 < n < 4$

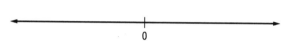

$m \leq 0$ or $m \geq 3$

Check whether the given value is a solution for the equation or inequality. Write *True* if the value is a solution and *False* if it is not.

4. Does $x = -1$? $5(x + 7) = 29 - x$

Does $b = -2$? $6b + 7 = 3b - 2$

5. Does $y = 3$? $-9(2y + 5) = -(y - 6)$

Does $a = 5$? $2(1 + a) = -6(3 - a)$

6. Does $c = 0$? $1 + 3c < -5c - 7$

Does $n = 4$? $6n - 7 \geq 1 + 4n$

7. Does $y = 2$? $-17 \leq 3y - 5 \leq 4$

Does $s = -2$? $1 > s - 2 > -4$

8. Does $b = -4$? $-9 + |2b - 10| = -7$

Does $m = -3$? $-6|6 + 4m| = -36$

Perform the indicated step. Do not solve.

9. Use the distributive property to eliminate parentheses:

$$-5(-a + 2) = -a$$

Clear the fractions by multiplying both sides by 8:

$$-\frac{3}{8}b + \frac{3}{2} = -\frac{3}{4}b$$

10. Multiply both sides by -1:

$-3x + 5 < -1$

Subtract 3 from all parts:

$5 < 2n + 3 < 11$

11. Multiply both sides by $\frac{5}{4}$:

$\frac{4}{5}y < \frac{12}{5}$

Divide both sides by -3:

$-3|5 - m| \leq -6$

Solve. Check your work.

12. $3x + 8 = 20$ $23 = 1 - 5a - 8$ $-4(7y + 8) - 6 = 46$

13. $n + 4 + 4n = -11$ $-y + 5y = 0$ $18 = -3(2 + 4x)$

14. $-8(b + 7) = 9 + 5b$ $4m - 11 = 3m - 5(m + 1)$ $3(2 + 6n) = 6(7n + 5)$

15. $-\frac{2}{3}m = 18$ $\frac{3}{4}a - 3 = 9$ $\frac{4}{3}x + 1 = 1 - 2x$

16. $-\frac{3}{5}b + \frac{19}{30} = -2b + \frac{2}{5}$ $\frac{1}{2}n + 1 = \frac{17}{50} + \frac{8}{5}n$ $-\frac{4}{5}y + \frac{1}{2}y = -\frac{7}{5} - \frac{3}{2}y$

17. $-2.3 = 0.7a + 0.8 - 5.9$ $1.31y + 3.2 + 7 = 11.51$ $6.3 + 6.4n = 5.8n + 1.5$

18. Three students solved the equation $-3 - 2x = -2(5 + x)$. Their work is shown below. Which solution is correct? For both incorrect solutions, underline the first line that contains a mistake.

A.
$$-3 - 2x = -2(5 + x)$$
$$-3 - 2x = -10 - 2x$$
$$-2x = -7 - 2x$$
$$-4x = -7$$
$$x = \frac{7}{4}$$

B.
$$-3 - 2x = -2(5 + x)$$
$$-3 - 2x = -10 - x$$
$$7 - 2x = -x$$
$$-x = -7$$
$$x = 7$$

C.
$$-3 - 2x = -2(5 + x)$$
$$-3 - 2x = -10 - 2x$$
$$7 - 2x = -2x$$
$$7 = 0$$
There is no solution: $\varnothing$.

Solve the following word problems.

19. On the first of three tests, Lynn scored 84. He scored one point more on the third test than he did on the second test. His average on the three tests was 91. What did he score on the second test?

20. There are two consecutive even integers. The first is 12 more than twice the second. What are the two integers?

21. Nita has a total of $6.30 in dimes and quarters. If she has twice as many dimes as quarters, how many dimes does she have?

22. $\angle A$ and $\angle B$ are supplementary. The measure of $\angle A$ is 2° less than six times the measure of $\angle B$. What is the measure of $\angle B$?

23. The length of a rectangle is 4 inches longer than twice its width. The perimeter of the rectangle is 62 inches. What is the width of the rectangle?

24. A landscaper is selling large plants for $4 each and medium plants for $2.50 each. A customer spent $93 on 30 plants. How many of each size plant did the customer buy?

25. In triangle XYZ, m$\angle X$ is 24° greater than four times the measure of $\angle Y$, and m$\angle Y$ is $\frac{2}{3}$ the measure of $\angle Z$. What is the measure of $\angle X$?

26. Three friends earned $147 on the weekend. Phil earned $6 less than 3 times Pat's earnings. Bob earned $5 less than twice Phil's earnings. How much did Bob earn?

27. In a parking lot, the number of compact spaces is 10 more than half the number of regular size spaces. If there are 30 compact spaces, how many parking spaces are there in all?

28. One angle of a triangle is 3 times greater than a second angle. The third angle is 60° less than the sum of the other two angles. What is the measure of each angle?

Solve and graph each inequality. Check your work.

29. $n - 2 \geq 10 + 4n$ $\qquad$ $-3m < 3 - 2m$ $\qquad$ $-2(x - 8) > -2(x + 7) + 3$

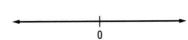 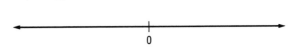 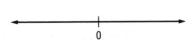

30. $11 - 3x \geq -5(x - 3)$ $\qquad$ $-25 - y \leq 6y - (1 - 5y)$ $\qquad$ $-2(3 + b) \leq -6(1 - 4b)$

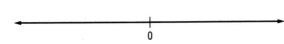 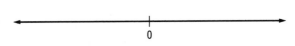

Solve and graph each inequality. Write your answer in interval notation.

31. $0 > 4x + 4 > -16$ $\qquad\qquad$ $2 + 7a < -54$ or $6 - 4a \leq -2$

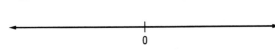 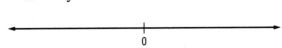

32. $7 \geq 7 - 2b \geq 5$ $\qquad\qquad$ $21 \leq 1 - 5y < 16$

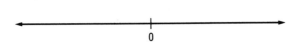 

33. $m - 3 > -5$ or $m + 7 < 9$ $\qquad\qquad$ $8 + 6n \geq 56$ or $n + 3 < -1$

 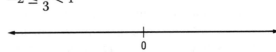

34. $19 \geq 6a + 1 \geq 13$ $\qquad\qquad$ $16 \leq -3y + 16$

35. $-4m - 1 \leq -29$ or $3m - 2 < -8$ $\qquad\qquad$ $-2 \leq \frac{n}{3} < 1$

Simplify.

36. $\dfrac{|-6 \cdot 2 + 3|}{3}$ 　　　　　　$5|3 - 7| - 12$ 　　　　　　$-|7 - 10| \cdot -2|-4|$

37. $|-9| + |2 - 8|$ 　　　　　　$-3|4 - 2| + 6$ 　　　　　　$-|4 - 10| - 2|-2 - 1|$

Solve.

38. $|4 + n| + 7 = 18$ 　　　　　$-3 + |a - 7| = -2$ 　　　　　$|-5x| + 2 = 27$

39. $-10|x - 9| = -110$ 　　　　$\dfrac{|b + 5|}{10} = 1$ 　　　　　$2 - \left|\dfrac{n}{8}\right| = 2$

40. $|-7 - 2b| - 7 = 8$ 　　　　$6|4x + 5| = 42$ 　　　　　$4|2 - 8a| = 8$

41. $4|4y + 10| - 5 = 3$ 　　　　$5|2 - 4n| + 2 = 12$ 　　　　$|8 - 5m| + 8 = 25$

42. $\dfrac{|9 - 2x|}{4} = 0$ 　　　　　$6|m + 9| = 54$ 　　　　　$\dfrac{|-5a - 2|}{2} = -5$

43. $|-10 + 4n| + 3 = 9$ 　　　　$7 - 4|2 + 2x| = -25$ 　　　　$1 - 2|8 + 4y| = 1$

Solve and graph each inequality. Then write your answer in interval notation.

44. $|b + 4| > 3$ 　　　　　　　　　　　$\left|\dfrac{x}{2}\right| > 2$

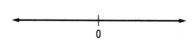　　　　　　　　　　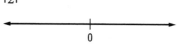

45. $|n + 3| < 7$ 　　　　　　　　　　　$|9a - 8| < -17$

46. $|-9x - 9| < 72$

$|8 + 3m| > -22$

(number line, 0 marked)

(number line, 0 marked)

47. $|6n - 3| - 6 < 33$

$\dfrac{|8y + 7|}{9} > -4$

(number line, 0 marked)

(number line, 0 marked)

Solve for the indicated variable.

48. Solve $A = \dfrac{1}{2}bh$ for b.

Solve $I = prt$ for p.

49. Solve $H = \dfrac{d^2N}{2.5}$ for N.

Solve $V = \dfrac{1}{3}bh$ for h.

50. Solve $R = 2h - \dfrac{1}{4}c$ for c.

Solve $A = \dfrac{a + b + c}{3}$ for b.

51. Solve $d = \dfrac{1}{a + b}$ for b.

Solve $V = \dfrac{3k}{t}$ for t.

52. Solve $S = 2lw + 2lh + 2wh$ for h.

Solve $A = \dfrac{1}{2}(b_1 + b_2)h$ for b_2.

Write each equation in terms of y.

53. $-5x + 5y = 10$

$6x - 3y = 9$

54. $\dfrac{x}{4} = \dfrac{y}{3}$

$4y - \dfrac{1}{2}x = 4$

LINEAR EQUATIONS AND GRAPHING

You have seen how the solution set to an equation or inequality with one variable can be shown on a number line. A number line has one dimension—length. When an equation or inequality has two variables, its graph must have two dimensions—length and width. In this section, you will learn how to graph the solution set of a linear equation on a coordinate plane.

Graphing on the Coordinate Plane

In this lesson, you will learn to

- Plot points on a coordinate plane.
- Graph an equation using a table of values.

The coordinate plane is formed by drawing two number lines that intersect at zero. You can describe any point on the grid by stating its location on the x-axis and then the y-axis. The location can be written in parentheses in the form (x, y).

In the grid at the right, point A is located at $(4, 3)$, and point B is located at $(-2, -5)$.

To plot a point or write its location, start at the origin $(0, 0)$ and count left or right for the x-coordinate, and then up or down for the y-coordinate.

How would you describe the location of point C on the grid?

The location is 6 to the right and 3 down, so the coordinates are written **(6, −3).**

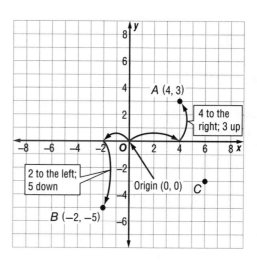

Write the coordinates of points D, E, F, and G.

1. D _____

2. E _____

3. F _____

4. G _____

Add points at the following locations:

5. J at $(-6, 3)$

6. K at $(0, -4)$

7. L at $(4, -2)$

8. M at $(3, 5)$

9. N at $(-2, -6)$

10. O at $(6, 0)$

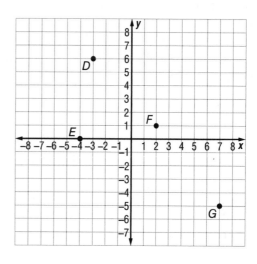

A linear equation has two variables, x and y. Any solution for a linear equation is written as an ordered pair—a value for each variable written in parentheses: (x, y).

To find a possible solution, choose a number for x and then solve for y. There will not be just one right answer. In fact, there are infinitely many solutions to a linear equation. You can organize the solutions in a table of values.

EXAMPLE 1 Create a table of values and a graph for the equation: $x + 2y = 7$

STEP 1 Choose three values for x and solve for y.

Let $x = 1$.
$$1 + 2y = 7$$
$$2y = 6$$
$$y = 3$$

Let $x = 3$.
$$3 + 2y = 7$$
$$2y = 4$$
$$y = 2$$

Let $x = 5$.
$$5 + 2y = 7$$
$$2y = 2$$
$$y = 1$$

STEP 2 Organize the solutions in a table of values.

Notice how the numbers in each column in the table form a pattern. Using the pattern, could you predict another pair of x- and y-values that would make the equation true?

x	y
1	3
3	2
5	1

STEP 3 Plot the points from the table onto the coordinate plane.

STEP 4 Draw a line through the points.

You can graph a line if you know two points, so why do the extra work to find a third point?

The third point will help you check your work. If you plot three points and all three don't line up, you have made a mistake and you need to check your work.

Every ordered pair that is a solution to the equation $x + 2y = 7$ will be located on the line in the graph. For example, the solution of the equation when 2 is used for x is shown at the right. Can you see that the point $(2, \frac{5}{2})$ is on the line?

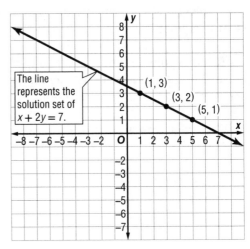

The line represents the solution set of $x + 2y = 7$.

$$x + 2y = 7$$
$$2 + 2y = 7$$
$$2y = 5$$
$$y = \frac{5}{2} \text{ or } 2\frac{1}{2}$$

Fraction values for x and y can be difficult to graph. Choose values for x carefully to avoid fractions.

EXAMPLE 2 Create a table of values for the equation: $y = \frac{1}{3}x + 1$

Notice that the variable x is being multiplied by $\frac{1}{3}$. If you choose multiples of 3 for x, then the value of y will always be an integer.

You could now graph the points $(3, 2)$, $(6, 3)$, and $(9, 4)$. A line going through those points would represent the solutions to the equation $y = \frac{1}{3}x + 1$.

x	y
3	2
6	3
9	4

For each equation, complete the table of values. Use the given values of x to find y.

11. $x - y = -2$

x	y
1	
2	
3	

$x - 3y = 12$

x	y
-3	
0	
3	

12. $3x - 4y = 0$

x	y
-4	
0	
4	

$3x + 2y = -8$

x	y
-4	
-2	
0	

13. $y = -5x + 2$

x	y
-2	
0	
2	

$-\frac{1}{4}x - 2 = y$

x	y
4	
8	
12	

Graph each equation by creating a table of values.

14. $y = \frac{3}{4}x - 1$

Choose your own values:

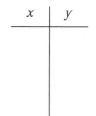

x	y

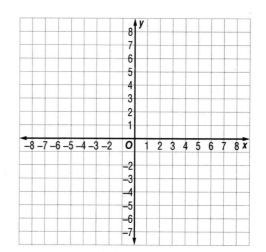

15. $3x - 7y = 21$

Choose your own values:

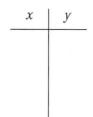

x	y

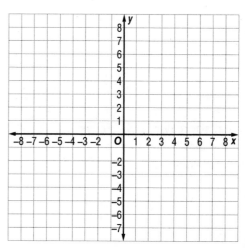

16. $2x + 5y = 10$

Choose your own values:

x	y

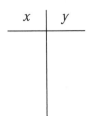

17. $y = \dfrac{2}{5}x + 1$

Choose your own values:

x	y

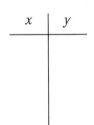

18. $5x - 4y = 20$

Choose your own values:

x	y

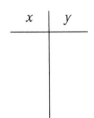

19. What strategy would you use to pick convenient values for the equation $x - 3y = -9$? Create a table of values using the strategy.

x	y

Slope and Intercepts

In this lesson, you will learn to

- Graph a line using x- and y-intercepts.
- Find the slope of a line.

Suppose you had to describe line m to a friend. Imagine that your friend has to try to draw the line based on your description alone. What could you say that would help your friend succeed?

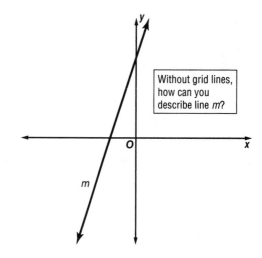

Without grid lines, how can you describe line m?

Your description might go something like this: *The line is very steep. It climbs very quickly going from left to right. It crosses the x-axis on the left side of 0, and it crosses the y-axis high up, much farther from 0.*

The description above describes the slope and the intercepts of the line.

- **Slope** is the measure of the steepness or incline of a line.
- The **x-intercept** is the point at which the line crosses the x-axis.
- The **y-intercept** is the point at which the line crosses the y-axis.

When you know the equation of a line, the intercepts are very easy to find.

The y-coordinate for any point on the x-axis must be 0. To find the x-intercept, let $y = 0$ and solve for x.

The x-coordinate for any point on the y-axis must be 0. To find the y-intercept, let $x = 0$ and solve for y.

You can use the two intercepts to graph an equation.

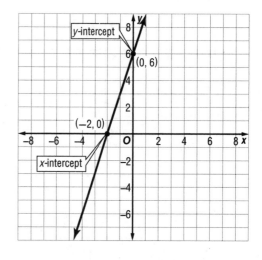

EXAMPLE 1   Use the intercept method to graph the equation $2x + y = 4$.

STEP 1   Solve for the intercepts.

Let $y = 0$.

$2x + y = 4$
$2x + 0 = 4$
$2x = 4$
$x = 2$

The x-intercept is $(2, 0)$.

Let $x = 0$.

$2x + y = 4$
$2(0) + y = 4$
$y = 4$

The y-intercept is $(0, 4)$.

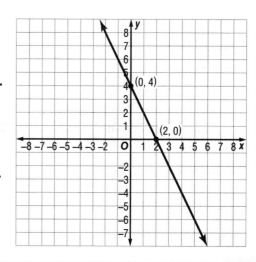

STEP 2   Plot the points and draw the line.

You can check your work with a third point. Look back at the graph on page 69. It appears that the line also passes through point $(1, 2)$. Try those values in the original equation for x and y, and see if the equation is true.

$$2(1) + 2 = 4$$
$$2 + 2 = 4$$
$$4 = 4 \text{ True}$$

 Caution: The intercept method for graphing a line is useful when the coordinates are integers. If one of the intercepts is a fraction, however, you may need to estimate the point's location. This could lead to an inaccurate graph.

Write the x- and y-intercepts for each equation. Then graph the equation.

1. $3x + y = 3$

x-intercept: _____

y-intercept: _____

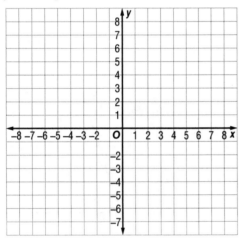

2. $y = -\dfrac{4}{5}x + 4$

x-intercept: _____

y-intercept: _____

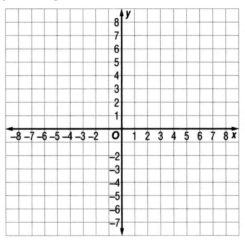

3. $y = -\dfrac{2}{3}x - 2$

x-intercept: _____

y-intercept: _____

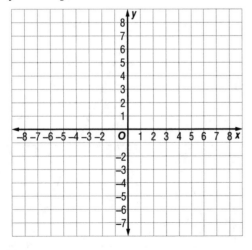

4. $4x + 2y = 8$

x-intercept: _____

y-intercept: _____

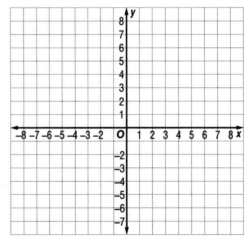

Look at the line at the right. Put your pencil on point *A* and draw to point *B*. As you follow the line, your pencil climbs higher as it moves farther to the right. You are moving in two directions at the same time—up and to the right.

Slope is a way to write both directions in one number. Slope is a ratio that compares two quantities: the **rise** (vertical change) and the **run** (horizontal change).

$$\text{slope} = \frac{\text{rise}}{\text{run}} = \frac{\text{change in } y}{\text{change in } x}$$

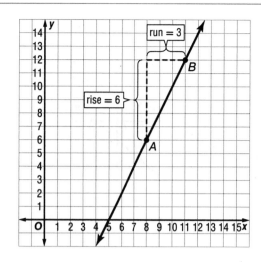

$$\text{slope of line } AB = \frac{\text{rise}}{\text{run}} = \frac{6}{3} = \mathbf{2}$$

If a line moves downward as it moves to the right, it has a negative slope.

EXAMPLE 2 Find the slope of a line that passes through points (−1, 4) and (7, 2).

STEP 1 Plot the points and draw the line.

STEP 2 Draw a vertical line to mark the rise. The line drops 2 spaces—a change in *y* of −2.

STEP 3 Draw a horizontal line to mark the run. The line moves 8 spaces to the right—a change in *x* of 8.

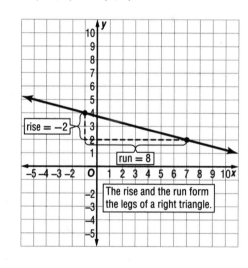

STEP 4 Write the ratio of rise to run and simplify.

$$\frac{\text{rise}}{\text{run}} = \frac{-2}{8} = -\frac{1}{4}$$

The slope of the line is $-\frac{1}{4}$.

Two cases of slope require special attention:

A horizontal line has no slope or steepness. It neither rises nor falls. The slope of a horizontal line is 0.

A vertical line has a run of 0. It doesn't move to the right or the left. The slope of a vertical line is undefined.

The slope of $y = 1$ is 0.

The slope of $x = 2$ is undefined.

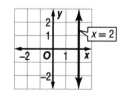

Why is the slope of a vertical line undefined? Remember that slope equals rise divided by run. The run of a vertical line is 0. Division by 0 is undefined in mathematics. So for a vertical line, the run equals 0 and the slope is undefined.

Find the slope of each line using the given points.

5.

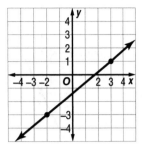

6.

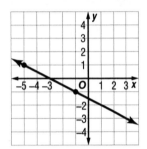

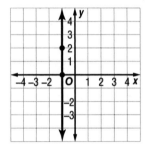

7.

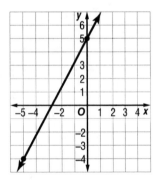

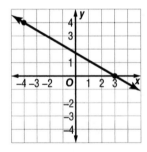

8.

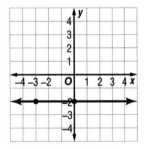

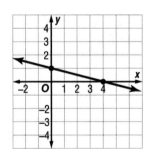

If you know one point on a line and the slope of the line, you can graph the line. This skill will help you check your work when graphing.

EXAMPLE 3 A line with a slope of $-\frac{2}{3}$ passes through point (−3, 1). Graph the line.

STEP 1 Plot the given point.

STEP 2 If the slope is $-\frac{2}{3}$, the rise is −2 and the run is 3. Since the rise is negative, count *down* 2 spaces. Then count 3 spaces to the right and plot the new point at (0, −1).

STEP 3 Draw a line through the two points.

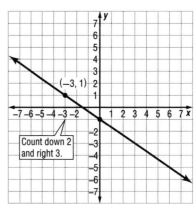

Use the slope and given point to graph each line.

9. Slope: $\frac{4}{5}$ Point: (0, −3)   Slope: $-\frac{1}{3}$ Point: (1, 4)

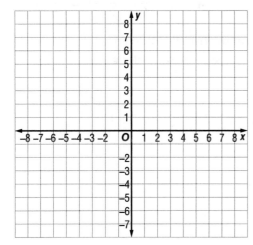 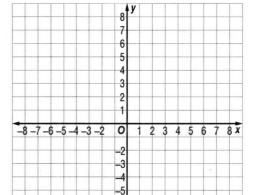

10. Slope: −3   Point: (−4, 4)   Slope: 2   Point: (−2, −5)
    (**Hint:** Use a rise of −3 over a run of 1.)

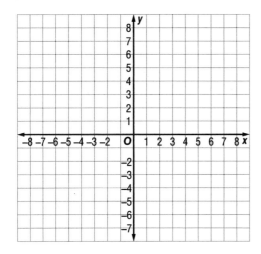 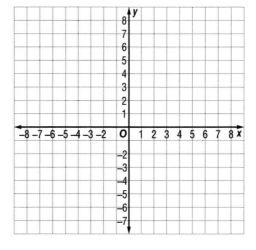

You can also find slope without a graph. As long as you have two points on a line, you can figure out the slope.

The formula for the **slope of a line** is $m = \frac{y_2 - y_1}{x_2 - x_1}$, where m is the slope and (x_1, y_1) and (x_2, y_2) are two points on the line.

EXAMPLE 4   Find the slope of a line that passes through points $(0, 5)$ and $(2, 6)$.

STEP 1   Let (x_1, y_1) represent $(0, 5)$ and (x_2, y_2) represent $(2, 6)$.

STEP 2   Substitute the values into the formula and solve.

$$m = \frac{y_2 - y_1}{x_2 - x_1} = \frac{6 - 5}{2 - 0} = \frac{1}{2}$$

The slope of the line is $\frac{1}{2}$.

You would get the same slope if you reversed the order of the points.

Let $(x_1, y_1) = (2, 6)$ and $(x_2, y_2) = (0, 5)$.

Why does the formula work?

$$m = \frac{y_2 - y_1}{x_2 - x_1} = \frac{5 - 6}{0 - 2} = \frac{-1}{-2} = \frac{1}{2}$$

Finding the difference in the y-values is the change in y, or the rise. The difference in the x-values is the change in x, or the run. The formula calculates the ratio of the rise to the run, which is the slope of the line.

 Caution: Keep in mind that division by zero is undefined. If $x_2 - x_1 = 0$, there is no run; the line must be vertical and its slope is undefined.

Find the slope of a line containing the following points.

11. $(-8, 3)$ and $(2, 5)$ $(-15, 12)$ and $(9, 0)$

12. $(-11, -13)$ and $(-8, -1)$ $(6, 2)$ and $(-5, 2)$

13. $(1, -7)$ and $(5, -2)$ $(1, 0)$ and $(1, -8)$

14. $(5, 4)$ and $(0, 4)$ $(7, -10)$ and $(5, -6)$

Finding the Equation of a Line

In this lesson, you will learn to

- Write an equation of a line given its slope and *y*-intercept.
- Write an equation of a line given its slope and a point on the line.
- Write an equation of a line given two points on the line.

You have learned three techniques for graphing an equation: creating a table of variables, finding and plotting the *x*- and *y*-intercepts, and using the slope and one point on the graph to determine the line. All three techniques work consistently. Knowing more than one technique is important because you now have choices when you approach a problem.

In this lesson, you are going to work in the opposite direction. You will be given a graph, or the information you could find on a graph, and you will be asked to write the equation of the line. Again, you will learn several techniques so that you can choose the best way to do a problem depending on the situation.

A linear equation can be written in several ways. Its most common form is **standard form,** which is written with the variables in order on the left side of the equation and the number constant on the right side. The coefficient of the term with *x* is a non-negative whole number. Mathematicians write equations in standard form to make it easier to compare equations.

EXAMPLES $2x + 3y = 8$ $x - 2y = 9$ $5x + y = -10$

Begin by comparing an equation to its graph.

Graph the equation $x - 3y = -6$.

Create a table of values and plot points.

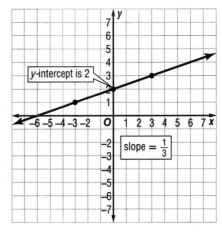

x	*y*
−3	1
0	2
3	3

Study the graph.
What could you say
about it?

The slope of the graph is $\frac{1}{3}$. The *x*-intercept is −6 and the *y*-intercept is 2.

The same equation can be written in a different form by solving for *y*. Using inverse operations, isolate the variable *y*. Watch what happens.

Solve for *y*: $x - 3y = -6$
$$-3y = -x - 6$$
$$y = \frac{1}{3}x + 2$$

The equation $y = \frac{1}{3}x + 2$ represents the same line as the equation in standard form. This new form, however, reveals a helpful pattern. In the new equation, the *x*-variable is being multiplied by the slope $\left(\frac{1}{3}\right)$ and the constant value is the *y*-intercept (2).

When solved for *y*, an equation is in **slope-intercept form.**

Slope-Intercept Form	$y = mx + b$, where m = slope and b = *y*-intercept

Using this form, it is simple to write the equation of a line if you know the slope and *y*-intercept.

EXAMPLE 1   Find the equation of a line with a slope of −3 and a *y*-intercept of −5.

STEP 1   Substitute the given values into the slope-intercept form.

$$y = mx + b = -3x + (-5)$$
$$y = -3x - 5$$

STEP 2   The equation is in slope-intercept form. Rewrite the equation in standard form by adding 3*x* to both sides of the equation.

$$3x + y = -5$$

You can also write the equation of a line by finding the slope and *y*-intercept on its graph.

EXAMPLE 2   Write the equation of the line shown on the graph below.

STEP 1   Find slope by counting rise over run.

STEP 2   Determine the *y*-intercept.

STEP 3   Substitute the values into slope-intercept form.

$$y = mx + b$$
$$y = 2x - 4$$

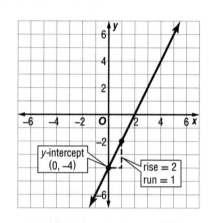

STEP 4   Rewrite the equation in standard form.

$$-2x + y = -4$$
$$-1(-2x + y) = -1(-4)$$
$$2x - y = 4$$

You already know how to graph a line given the slope and one point on the line. Because slope-intercept form reveals the slope and one point (the *y*-intercept), you can use it to graph the equation. Just count the rise and run from the *y*-intercept.

EXAMPLE 3   Graph $x + 4y = 12$.

STEP 1   Write the equation in slope-intercept form.

$$x + 4y = 12$$
$$y = -\frac{1}{4}x + 3$$

STEP 2   Plot the *y*-intercept: (0, 3)

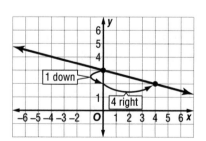

STEP 3   Think of the slope as $\frac{\text{rise}}{\text{run}} = \frac{-1}{4}$. Count 1 down and 4 to the right. Plot another point, and draw a line through the points.

Write an equation for a line given the slope (*m*) and *y*-intercept (*b*). Write the equation first in slope-intercept form and then in standard form.

	Slope-Intercept Form	Standard Form

1. $m = -3 \quad b = 4$ **a.** _____   **b.** _____

2. $m = 1 \quad b = -3$ **a.** _____   **b.** _____

3. $m = 2 \quad b = -5$ **a.** _____   **b.** _____

4. $m = -1 \quad b = 0$ **a.** _____   **b.** _____

5. $m = \frac{2}{3} \quad b = 3$ **a.** _____   **b.** _____

6. $m = -\frac{2}{5} \quad b = 2$ **a.** _____   **b.** _____

7. $m = \frac{7}{4} \quad b = -1$ **a.** _____   **b.** _____

8. $m = -\frac{3}{4} \quad b = -2$ **a.** _____   **b.** _____

For each graph, write the equation of the line in slope-intercept form.

9.

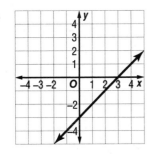

10.

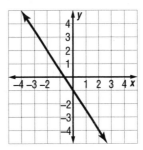

11.

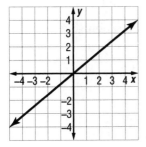

12.

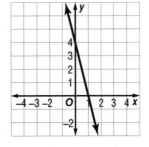

Suppose you know that a line with slope -3 passes through point $(1, -2)$. Is there any way to write the equation using only this information?

What would you do? You don't know the y-intercept so you can't use slope-intercept form. You could draw a graph of the line so you could see the y-intercept. If the y-intercept is a fraction, however, you would have to estimate, and you would not be able to write the exact equation.

Try to use what you know to develop a way to solve this problem. You know the slope, and you know the formula for how to find slope. You also know a point on the line. You can substitute the facts you know for the variables in the slope formula.

EXAMPLE 4 Write the equation of a line with a slope of -3 that passes through $(1, -2)$.

STEP 1 Substitute the facts into the slope formula. Replace m with the given slope in the slope formula. Let (x_1, y_1) be the point $(1, -2)$. You can change (x_2, y_2) to (x, y) since they are the only variables left.

$$m = \frac{y_2 - y_1}{x_2 - x_1}$$

$$-3 = \frac{y - (-2)}{x - 1}$$

Now solve for y.

STEP 2 Multiply both sides by $x - 1$.

STEP 3 Subtract 2 from both sides.

STEP 4 Exchange the sides.

$$-3(x - 1) = y + 2$$
$$-3x + 3 = y + 2$$
$$-3x + 1 = y$$
$$y = -3x + 1$$

You can see from the slope-intercept form that the line has slope -3. You can do a check to make sure the point $(1, -2)$ is on the line.

CHECK Substitute $(1, -2)$ for x and y.

$(1, -2)$ is a solution to the equation; this means it is a point on the line. You can rewrite the equation of the line in standard form as $3x + y = 1$.

$$y = -3x + 1$$
$$-2 = -3(1) + 1$$
$$-2 = -3 + 1$$
$$-2 = -2 \text{ True}$$

You can now write an equation of a line when only the slope and one point on the line are given. The form of the equation used in Example 4 is called **point-slope.** Do you see how it gets its name? Use it when you know only a point and the slope of a line.

Point-Slope Form $y - y_1 = m(x - x_1)$, where $m =$ slope and (x_1, y_1) is a known point.

Being good at math requires some memorization. To be good at algebra and graphing, you should master the slope formula, the slope-intercept form, and the point-slope form. Practice writing them. Think about each variable and what it represents. Look for the ways in which the forms are the same and how they are different.

In the next example, using the point-slope form will save you a lot of time and many steps. It works because the steps are already incorporated into the form. When you use the form, you are performing all the steps at once.

EXAMPLE 5   Write the equation of a line with a slope of 7 that passes through (3, 4). Write your answer in slope-intercept form.

STEP 1   Substitute the facts into point-slope form. Let $m = 7$, $x_1 = 3$, and $y_1 = 4$.

STEP 2   Solve for y.

$$y - y_1 = m(x - x_1)$$
$$y - 4 = 7(x - 3)$$
$$y - 4 = 7x - 21$$
$$y = 7x - 17$$

That was much easier, wasn't it?

The slope and y-intercept of a line can often be fractions. When you write an equation in standard form, you should clear the fractions by multiplying by the lowest common denominator (LCD).

EXAMPLE 6   Write the equation of a line with a slope of $-\frac{3}{7}$ that passes through (5, −1). Write the equation in standard form.

STEP 1   Use point-slope form to first write the equation in slope-intercept form. Let $m = -\frac{3}{7}$, $x_1 = 5$, and $y_1 = -1$.

Notice that the y-intercept is $\frac{8}{7}$. You wouldn't have been able to write this equation just by looking at its graph.

STEP 2   Now multiply both sides by 7, and rewrite the equation in standard form. The equation is $3x + 7y = 8$.

$$y - y_1 = m(x - x_1)$$
$$y - (-1) = -\frac{3}{7}(x - 5)$$
$$y + 1 = -\frac{3}{7}x + \frac{15}{7}$$
$$y = -\frac{3}{7}x + \frac{8}{7}$$
$$7y = -3x + 8$$
$$3x + 7y = 8$$

Write an equation for each line with the given point and slope.
Write the equation in slope-intercept form.

13.   (1, −3), slope = −8   (−1, −2), slope = 3

14.   (1, −5), slope = −7   (1, 3), slope = 4

15.   (−4, 3), slope = $\frac{1}{2}$ (2, −2), slope = 0

16.   (4, −3), slope = $\frac{1}{4}$ (−5, −1), slope = $\frac{6}{5}$

17.   (2, −5), slope = $-\frac{10}{3}$ (5, 4), slope = $\frac{1}{2}$

Write an equation for each line with the given point and slope. Write the equation in standard form.

18. $(5, -1)$, slope $= \dfrac{2}{5}$ $(5, 4)$, slope $= \dfrac{9}{5}$

19. $(-4, 3)$, slope $= -\dfrac{7}{4}$ $(-3, -2)$, slope $= \dfrac{7}{3}$

20. $(-4, -5)$, slope $= \dfrac{5}{2}$ $(-3, 2)$, slope $= -\dfrac{2}{3}$

21. $(-3, -3)$, slope $= -\dfrac{1}{3}$ $(-5, 4)$, slope $= -\dfrac{2}{5}$

22. $(3, 4)$, slope $= \dfrac{5}{3}$ $(5, 5)$, slope $= \dfrac{1}{5}$

If you know only two points on a line, you can still use the point-slope form to write an equation. First, solve for slope. Then use the point-slope form with one of the points. Choose the point that will make calculations as easy as possible.

EXAMPLE 7   Write an equation for a line that passes through $(4, -4)$ and $(3, 0)$. Write the equation in slope-intercept form.

STEP 1   Use the slope formula. Substitute $(3, 0)$ for (x_1, y_1) and $(4, -4)$ for (x_2, y_2).

The slope is -4.

$$m = \frac{y_2 - y_1}{x_2 - x_1}$$
$$= \frac{-4 - 0}{4 - 3}$$
$$= \frac{-4}{1}$$
$$= -4$$

STEP 2   Use the point-slope form with -4 for slope. This solution uses $(3, 0)$ for (x_1, y_1).

$$y - y_1 = m(x - x_1)$$
$$y - 0 = -4(x - 3)$$
$$y = -4x + 12$$

Hint: It doesn't matter which point you use for (x_1, y_1) and (x_2, y_2). It is very important, however, that you use only the y-coordinates for y and only the x-coordinates for x. It's easy to switch them accidentally. You may find it helpful to write labels above the coordinates so that it is easy to remember which variables are assigned to each value. It is also very important that you use the x-coordinate and the y-coordinate values from the same point.

In the next example, you are given a graph with two marked points.

EXAMPLE 8 Write an equation in standard form for the line shown on the graph.

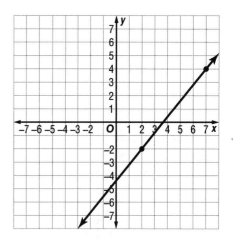

STEP 1 Determine the coordinates of the points. They are (7, 4) and (2, −2).

STEP 2 Find the slope. Use the formula or count spaces to find rise over run.

Let $(7, 4) = (x_2, y_2)$ and $(2, -2) = (x_1, y_1)$.

$$m = \frac{y_2 - y_1}{x_2 - x_1} = \frac{4 - (-2)}{7 - 2} = \frac{6}{5}$$

STEP 3 Use the point-slope form.

$$y - y_1 = m(x - x_1)$$

$$y - (-2) = \frac{6}{5}(x - 2)$$

$$y + 2 = \frac{6}{5}x - \frac{12}{5}$$

$$y = \frac{6}{5}x - \frac{22}{5}$$

STEP 4 Write the equation in standard form. To clear the fractions, multiply both sides by 5.

$$y = \frac{6}{5}x - \frac{22}{5}$$

$$5y = 6x - 22$$

$$-6x + 5y = -22$$

$$-1(-6x + 5y) = -1(-22)$$

$$6x - 5y = 22$$

In the next section of exercises, you will be able to choose any method for writing the equation of a line. For your convenience, the forms you have learned are listed below. Take time to master them. You will use them throughout your study of algebra.

Slope (m): $m = \frac{y_2 - y_1}{x_2 - x_1}$, where (x_1, y_1) and (x_2, y_2) are points on the line.

Slope-Intercept Form: $y = mx + b$, where m = slope and b = y-intercept.

Point-Slope Form: $y - y_1 = m(x - x_1)$, where m = slope and (x_1, y_1) is a known point.

Write the equation of the line that contains the following points. Write your answer in slope-intercept form.

23. (2, 0) and (0, 5)   (1, 1) and (−3, −1)

24. (0, 1) and (4, −2)   (1, −2) and (−2, −3)

25. $(3, -4)$ and $(-3, 3)$ $(3, -5)$ and $(0, 5)$

26. $(4, 4)$ and $(0, 0)$ $(4, 0)$ and $(-4, -3)$

Write the equation of each graphed line. Write your answer in standard form.

27.

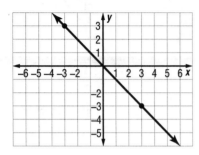

28.

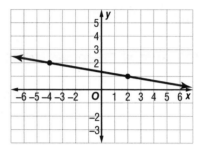

29.

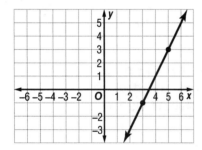

30.

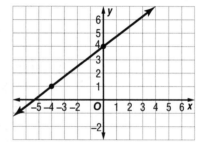

31.

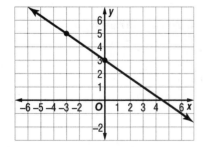

32.

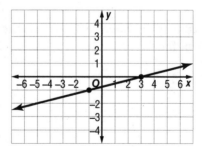

Parallel and Perpendicular Lines

In this lesson, you will learn to

- Graph and write equations for parallel and perpendicular lines.

Examine the three different lines drawn on the same coordinate plane below.

Lines *a* and *b* are parallel, and line *c* is perpendicular to both *a* and *b*.

Parallel lines lie in the same plane and never intersect. **Perpendicular lines** intersect to form 90° angles.

You can use algebra to add to your understanding of geometry. First write the equation of each line. Because you can easily see the *y*-intercept of all three lines, just use the slope-intercept form.

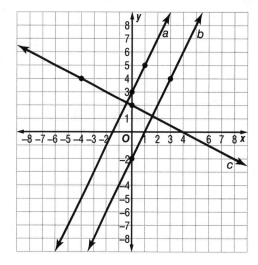

line *a*   slope $= 2$
       y-intercept $= (0, 3)$
       $y = 2x + 3$

line *b*   slope $= 2$
       y-intercept $= (0, -2)$
       $y = 2x - 2$

line *c*   slope $= -\frac{1}{2}$
       y-intercept $= (0, 2)$
       $y = -\frac{1}{2}x + 2$

Do you notice any patterns? The parallel lines have the same slope. This makes sense because lines that have the same slope have exactly the same slant. If they pass through different points, they will never intersect.

> **Key Idea:** Two lines are parallel if they have the same slope.

Examine the results again. How is the slope of line *c* different from the slopes of *a* and *b*? The slope of line *c* is negative while *a* and *b* have positive slopes. Also, notice that $\frac{1}{2}$ is the reciprocal of 2. So $-\frac{1}{2}$ (the slope of line *c*) is the negative reciprocal of 2 (the slope of lines *a* and *b*).

> Remember: Find the reciprocal of a number by dividing 1 by that number. When you multiply reciprocals, the product is 1.
>
> The reciprocal of *a* is $\frac{1}{a}$.
>
> The negative reciprocal of *a* is $-\frac{1}{a}$.

> **Key Idea:** Two lines are perpendicular if one slope is the negative reciprocal of the other.

EXAMPLE 1 Determine whether the graphs of the equations $-7x + 2y = 8$ and $14x - 4y = 3$ are parallel.

STEP 1 Rewrite both equations in slope-intercept form.

$$-7x + 2y = 8 \qquad\qquad 14x - 4y = 3$$
$$2y = 7x + 8 \qquad\qquad -4y = -14x + 3$$
$$y = \frac{7}{2}x + 4 \qquad\qquad y = \frac{-14}{-4}x - \frac{3}{4}$$

STEP 2 Compare the slope of both equations.

$$y = \frac{7}{2}x - \frac{3}{4}$$

Both lines have a slope of $\frac{7}{2}$. Since the slopes are equal, **the lines are parallel.**

EXAMPLE 2 Determine whether the graphs of the equations $-2x + 3y = 9$ and $3x + 2y = 2$ are perpendicular.

STEP 1 Rewrite both equations in slope-intercept form.

$$-2x + 3y = 9 \qquad\qquad 3x + 2y = 2$$
$$3y = 2x + 9 \qquad\qquad 2y = -3x + 2$$
$$y = \frac{2}{3}x + 3 \qquad\qquad y = -\frac{3}{2}x + 1$$

STEP 2 Compare the slopes.

The slope $-\frac{3}{2}$ is the negative reciprocal of $\frac{2}{3}$, so **the lines are perpendicular.**

Decide whether each pair of lines is parallel, perpendicular, or neither.

1. $y = -\frac{1}{2}x + 5$ $8x + 2y = 0$ $-x + 2y = -1$
    $x + 2y = 8$ $8x + y = -28$ $2x + y = -4$

2. $3x + 7y = 35$ $y = \frac{1}{6}x$ $4x + 3y = 15$
    $-7x + 3y = 9$ $x - 6y = 28$ $-x + 4y = 5$

3. $3y = x$ $-7x + 5y = 20$ $y = -\frac{2}{5}x - 4$
    $y = -3x + 3$ $y = \frac{5}{7}x - 1$ $2x + 5y = -5$

Using your understanding of lines and graphing, you can determine the equation of a parallel or perpendicular line that passes through a particular point.

EXAMPLE 3 Write an equation for a line that passes through $(-3, 4)$ and is parallel to the graph of the line $y = 2x - 4$.

STEP 1 Determine the slope of the given equation. The equation $y = 2x - 4$ is already in slope-intercept form; the slope is 2.

$$y - y_1 = m(x - x_1)$$
$$y - 4 = 2(x - (-3))$$

STEP 2 Use the given point and the point-slope form to write the new equation.

$$y - 4 = 2(x + 3)$$
$$y - 4 = 2x + 6$$
$$y = 2x + 10$$

The equations $y = 2x - 4$ and $y = 2x + 10$ must be parallel because they have the same slope.

You can use the given point in the new equation to check your work.

Knowing the slope of the lines is the key for working with parallel or perpendicular lines. Unless you are given a graph, you must first write the equations in slope-intercept form. Successful students do not skip this step.

EXAMPLE 4 Write an equation for a line that passes through $(3, -2)$ and is perpendicular to the graph of the line $3x + y = -2$.

STEP 1 Determine the slope of the given equation. Write the equation in slope-intercept form. The slope is -3.

$$3x + y = -2$$
$$y = -3x - 2$$

STEP 2 Find the negative reciprocal. To invert -3, think of it as $-\frac{3}{1}$. Invert the fraction and change the sign from negative to positive.

$$-3 \longrightarrow \frac{1}{3}$$

STEP 3 Use the given point $(3, -2)$, the slope $\left(\frac{1}{3}\right)$, and the point-slope form to write the new equation.

$$y - y_1 = m(x - x_1)$$
$$y - (-2) = \frac{1}{3}(x - 3)$$
$$y + 2 = \frac{1}{3}x - 1$$
$$y = \frac{1}{3}x - 3$$

The equations $3x + y = -2$ and $y = \frac{1}{3}x - 3$ must be perpendicular because their slopes are negative reciprocals.

Check the given point in the new equation.

$$y = \frac{1}{3}x - 3$$
$$-2 = \frac{1}{3} \cdot 3 - 3$$
$$-2 = 1 - 3$$
$$-2 = -2 \text{ True}$$

It is always a good idea to check your work in some way. When problems contain many steps, there are many opportunities for mistakes to occur. Lining up the equal sign as you write steps is a good way to keep yourself organized. When you can, check your work by substituting given points into the equation or by graphing the lines on a calculator.

Write equations as indicated.

4. Write an equation for a line that passes through point $(5, -3)$ and is parallel to $y = -x - 4$.

   Write an equation for a line that passes through point $(-1, 2)$ and is parallel to $y = 6x + 3$.

5. Write an equation for a line that passes through point $(2, -4)$ and is parallel to $y = -3x - 1$.

   Write an equation for a line that passes through point $(5, 5)$ and is parallel to $y = \frac{6}{5}x - 3$.

6. Write an equation for a line that passes through point $(2, 0)$ and is perpendicular to $y = -2x + 3$.

Write an equation for a line that passes through point $(-2, 2)$ and is perpendicular to $y = -\frac{2}{3}x + 1$.

7. Write an equation for a line that passes through point $(-2, 4)$ and is perpendicular to $y = x + 5$.

Write an equation for a line that passes through point $(-5, -5)$ and is perpendicular to $y = -x$.

8. Write an equation for a line that passes through point $(-2, -5)$ and is parallel to $y = \frac{3}{4}x + 3$.

Write an equation for a line that passes through point $(4, 0)$ and is perpendicular to $y = -\frac{3}{5}x - 3$.

For problems 9 and 10, write your answers in standard form.

9. Write an equation for a line that passes through point $(1, 4)$ and is parallel to $-6x + y = 2$.

Write an equation for a line that passes through point $(4, 0)$ and is perpendicular to $-x + y = 1$.

10. Write an equation for a line that passes through point $(-3, -2)$ and is parallel to $-5x + 3y = -9$.

Write an equation for a line that passes through point $(-5, 1)$ and is perpendicular to $4x + y = 0$.

11. **Challenge** Using the line segment AB on the graph at the right, write the equations for three other lines that together with AB would form a rectangle with a vertex at point C.

   Graph the equations on the coordinate plane. One of the lines is already drawn.

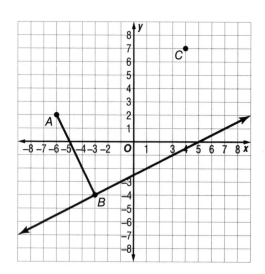

12. If you draw the diagonals of the rectangle $ABCD$ in problem 11, will the diagonals be perpendicular? Explain your thinking.

Problem Solving on the Coordinate Plane

In this lesson, you will learn to

- Find the midpoint of a segment on the coordinate plane.
- Find the distance between two points on the coordinate plane.
- Apply the Pythagorean theorem to the coordinate plane.

Sometimes algebraic formulas look more complicated than they really are. For example, finding the midpoint of a segment is something you can probably do using common sense.

Look at the number line. Which point on the line is the midpoint?

Keep in mind that the midpoint is the point that is an equal distance from both ends—in other words, the midpoint is the halfway point.

You can tell by just looking that the midpoint is **6.** In this example, the midpoint is 4 units from each endpoint.

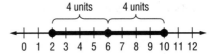

Without the number line, could you still find the midpoint? Using the idea that the midpoint is the halfway point, you could add the endpoints and divide by 2.

$$\frac{2 + 10}{2} = \frac{12}{2} = 6$$

This same strategy can be applied to find the endpoint of a segment on the coordinate plane. First look at the geometric solution.

EXAMPLE 1 Find the midpoint of segment *AB* with endpoints at (–2, –4) and (6, 6).

STEP 1 Plot the endpoints and draw the segment.

STEP 2 Find the middle of the vertical distance.

STEP 3 Find the middle of the horizontal distance.

STEP 4 Find the point where the two halfway points intersect. The midpoint is (2, 1).

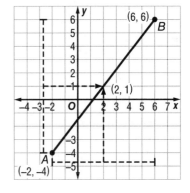

Now get ready to solve the problem algebraically. You will find the average of the *x*-values and the average of the *y*-values of the two endpoints. The two averages become the coordinates of the midpoint.

Here is the formula, but don't be intimidated by it. It's really a simple process.

For endpoints (x_1, y_1) and (x_2, y_2), the midpoint of the segment is equal to:

$$\left(\frac{x_1 + x_2}{2}, \frac{y_1 + y_2}{2}\right)$$

EXAMPLE 2   Find the midpoint of segment AB with endpoints at $(-2, -4)$ and $(6, 6)$.

Use the formula.
Let $(-2, -4) = (x_1, y_1)$ and $(6, 6) = (x_2, y_2)$.

The midpoint is **(2, 1)**.

$$\left(\frac{x_1 + x_2}{2}, \frac{y_1 + y_2}{2}\right) = \left(\frac{-2 + 6}{2}, \frac{-4 + 6}{2}\right)$$
$$= \left(\frac{4}{2}, \frac{2}{2}\right)$$
$$= \textbf{(2, 1)}$$

For each pair of points, find the midpoint. Write improper fractions as mixed numbers. For example, write $\frac{7}{2}$ as $3\frac{1}{2}$.

1. $(1, 2)$ and $(-9, 10)$ $(-6, 2)$ and $(-8, -8)$

2. $(4, -4)$ and $(8, 10)$ $(8, 0)$ and $(8, 6)$

3. $(-9, 4)$ and $(-2, -6)$ $(1, 2)$ and $(-1, -3)$

4. $(4, 3)$ and $(8, -3)$ $(-9, 5)$ and $(-8, -2)$

5. Find the midpoint of each leg of triangle PQR.

    midpoint of side PQ _____

    midpoint of side QR _____

    midpoint of side PR _____

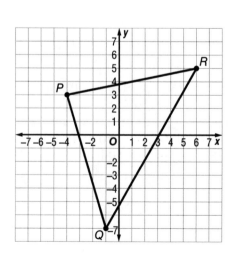

Another important problem-solving tool is the ability to find the distance between points on the coordinate plane.

EXAMPLE 3   Find the distance between points (5, 6) and (5, −2).

After plotting the two points on a coordinate plane, you can see that you need to measure only the vertical distance, which can be done by counting.

You can also, however, solve the problem using algebra. Subtract one *y*-coordinate from the other.

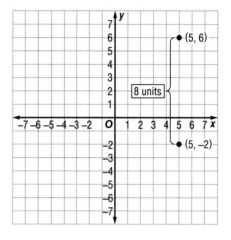

$|-2 - 6| = |-8| = $ **8 units** or
$|6 - (-2)| = |8| = $ **8 units**

Can you see why absolute value is used to solve distance problems? It isn't possible for distance to be negative, so the absolute value symbol ensures the answer will be positive. Using absolute value, you can subtract points in any order and get the same answer.

EXAMPLE 4   Find the distance between points (−5, −4) and (7, −4).

STEP 1   Plot the points. Notice that you need to measure only the horizontal distance.

STEP 2   Subtract one *x*-coordinate from the other. Again, the order in which you subtract the points doesn't matter.

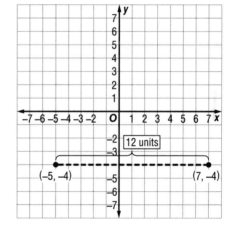

$|-5 - 7| = |-12| = $ **12 units** or
$|7 - (-5)| = |12| = $ **12 units**

So far, you have been able to solve distance problems with minimal calculations. Distance problems become more difficult when the points lie on a diagonal line.

Suppose you need to find the distance between the points shown on the graph.

You can see that there isn't an easy way to count out the distance. You have to take both horizontal and vertical change into account.

How can you measure a diagonal line?

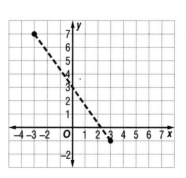

Here's a review of a formula you learned in geometry for working with right triangles.

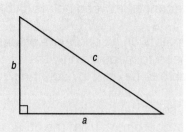

Pythagorean Theorem

In a right triangle, the sum of the squares of the lengths of the legs is equal to the square of the length of the hypotenuse.

$$a^2 + b^2 = c^2$$

Remember that the hypotenuse, represented by the variable c, is the side of the triangle that is directly opposite the right angle. The legs a and b form the right angle. Now use these relationships to solve the problem.

EXAMPLE 5   What is the distance between points $(-3, 7)$ and $(3, -1)$?

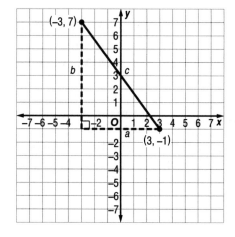

STEP 1   Think of the distance you need to find as the hypotenuse of a right triangle.

STEP 2   Find the lengths of the two legs of the triangle. Subtract the x-coordinates to find the horizontal leg, and subtract the y-coordinates to find the vertical leg.

horizontal: $a = |-3 - 3| = |-6| = 6$

vertical: $b = |7 - (-1)| = |8| = 8$

The two legs measure 6 and 8 units.

STEP 3   Use the Pythagorean theorem and solve for the hypotenuse c. Plug in values for a and b.

$$a^2 + b^2 = c^2$$
$$6^2 + 8^2 = c^2$$
$$36 + 64 = c^2$$

Simplify.

$$100 = c^2$$

Take the square root of both sides.

$$\sqrt{100} = \sqrt{c^2}$$
$$c = 10$$

> Distance cannot be negative, so you need only the positive root.

The distance between the points is **10 units.**

As you have seen before, algebra can be used to combine all these steps into one formula.

Distance Between Two Points $d = \sqrt{(x_2 - x_1)^2 + (y_2 - y_1)^2}$

Take a minute to study the parts of the formula. First you find the difference in x-coordinates and square the result; then you find the difference in y-coordinates and square the result. You add the squared numbers, and then find the square root of the sum. These are the exact same steps you used in Example 5 above.

Solve the next example using the distance formula.

EXAMPLE 6   What is the distance between points $(-2, -8)$ and $(0, -4)$?

STEP 1   Assign the points to the variables. Let $(-2, -8)$ be represented by (x_1, y_1), and let $(0, -4)$ be represented by (x_2, y_2).

STEP 2   Substitute the values into the formula and solve.

Remember that subtracting a negative is the same as adding.

Be sure to simplify your answer. To review simplifying radicals, see page 10.

The distance between the points is $2\sqrt{5}$ units.

$$d = \sqrt{(x_2 - x_1)^2 + (y_2 - y_1)^2}$$
$$= \sqrt{(0 - (-2))^2 + (-4 - (-8))^2}$$
$$= \sqrt{(0 + 2)^2 + (-4 + 8)^2}$$
$$= \sqrt{(2)^2 + (4)^2}$$
$$= \sqrt{4 + 16}$$
$$= \sqrt{20}$$
$$= 2\sqrt{5}$$

Graph each pair of points and find the distance between them. Use the Pythagorean theorem when necessary.

6. $(-2, -2)$ and $(2, -5)$

7. $(-5, -1)$ and $(-5, 8)$

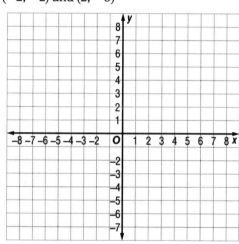

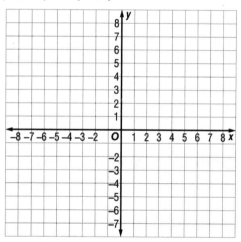

8. $(-7, 1)$ and $(-1, 7)$

9. $(5, 1)$ and $(-4, -2)$

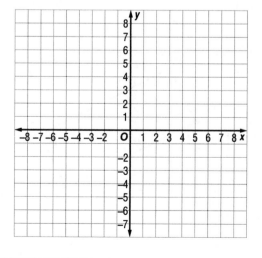

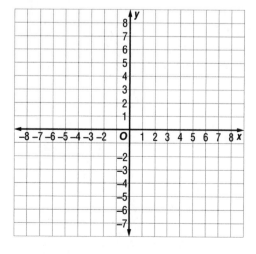

Use the distance formula to determine the distance between each pair of points.
Write radicals in simplified form.

10. $(5, 1)$ and $(-3, 5)$ $(3, 4)$ and $(3, -8)$

11. $(8, -4)$ and $(-3, 7)$ $(-7, -1)$ and $(-1, 5)$

12. $(6, -6)$ and $(-3, 0)$ $(4, 6)$ and $(7, 3)$

13. $(8, -2)$ and $(-1, -2)$ $(-8, 4)$ and $(0, 8)$

14. $(8, 6)$ and $(3, 1)$ $(7, 0)$ and $(-5, -4)$

15. Triangle *ABC* has vertices at *A*(7, 6), *B*(−8, 1), and *C*(−2, −7). Graph the triangle on the coordinate plane. Is triangle *ABC* an isosceles triangle? Explain your answer.

    (**Hint:** An isosceles triangle has two sides of equal length.)

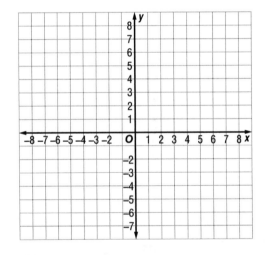

16. Rectangle *JKLM* is shown on the coordinate plane with point *K* at $(-3, 5)$. Find the perimeter of the rectangle.

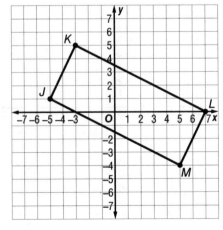

Linear Equations and Graphing Review

Solve the problems below. When you finish, check your answers at the back
of the book, and correct any errors.

For each equation, complete the table of values.

1. $2x - y = -5$

x	y
1	
2	
3	

$2x - 3y = 6$

x	y
-3	
0	
3	

Find the x- and y-intercepts for each equation. Then graph the equation on the coordinate plane.

2. $5x + 2y = 10$

$y = \frac{3}{4}x - 3$

x-intercept: _____ y-intercept: _____

x-intercept: _____ y-intercept: _____

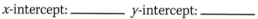
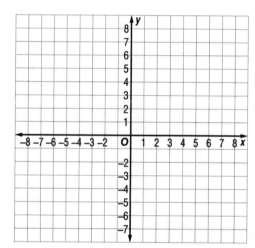

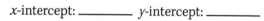
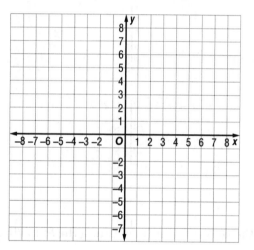

Use the given points to find the slope of each line on the coordinate plane.

3.

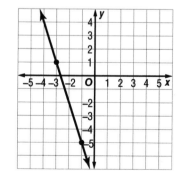

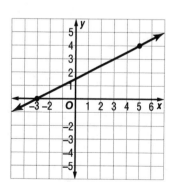

Use the slope and given point to graph the line.

4. Slope: $\frac{5}{4}$ Point: $(3, -2)$

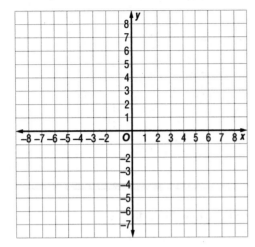

   Slope: $-\frac{2}{3}$ Point: $(-3, 4)$

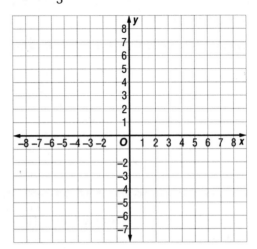

5. Slope: -1 Point: $(-5, 1)$

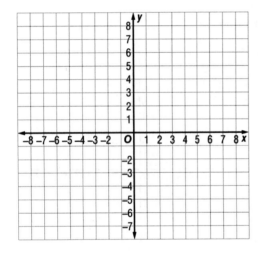

   Slope: 0 Point: $(0, 2)$

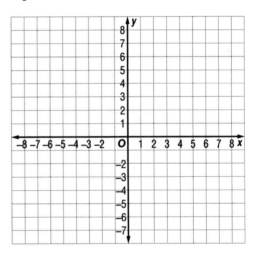

Find the slope of a line containing the following points.

6. $(-6, 3)$ and $(-7, -1)$ $(-4, 8)$ and $(8, -20)$

7. $(-18, 7)$ and $(-13, 9)$ $(3, 10)$ and $(3, -6)$

8. $(-6, 8)$ and $(0, 5)$ $(5, -4)$ and $(-5, -7)$

9. $(8, 10)$ and $(16, -6)$ $(4, -6)$ and $(9, -6)$

Given the *y*-intercept and slope, write the equation of the line in standard form.

10. $(0, -5)$, slope $= 6$ $(0, -3)$, slope $= \frac{1}{4}$

11. $(0, 3)$, slope $= \frac{2}{3}$ $(0, 0)$, slope $= -\frac{3}{4}$

12. $(0, -2)$, slope $= \frac{3}{2}$ $(0, -3)$, slope $= \frac{1}{5}$

Given the slope and a point on the line, write the equation of the line in slope-intercept form.

13. $(-5, -1)$, slope $= 1$ $(4, 1)$, slope $= -\frac{1}{4}$

14. $(-3, 3)$, slope $= -\frac{5}{3}$ $(1, -5)$, slope $= 0$

15. $(3, -1)$, slope $=$ undefined $(-4, 1)$, slope $= -\frac{5}{6}$

For each set of coordinates below, write the equation of a line that passes through both points. Write your answer in standard form.

16. $(0, 5)$ and $(4, 4)$ $(4, -3)$ and $(0, 3)$

17. $(2, 1)$ and $(0, -4)$ $(0, 3)$ and $(-2, -4)$

18. $(-5, 4)$ and $(-2, 0)$ $(-3, -4)$ and $(4, -2)$

19. $(-2, 2)$ and $(-5, -2)$ $(-3, -5)$ and $(-5, -1)$

Write the equation of each graphed line. Write your answer in standard form.

20.

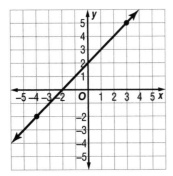
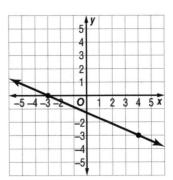

21.

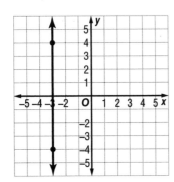
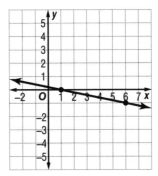

22.

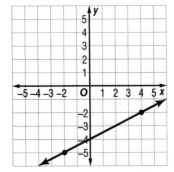
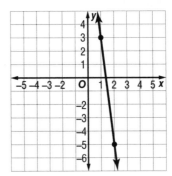

Decide whether each pair of lines is parallel, perpendicular, or neither.

23. $2x - y = 4$
$x - \frac{1}{2}y = \frac{5}{2}$

$y = -4x + 3$
$3y = 4x + 12$

$6y = x + 30$
$y = -6x + 4$

24. $-8x + y = 5$
$y = \frac{1}{8}x$

$-5x + 2y = -6$
$5x - 2y = -10$

$-3x + 2y = -6$
$2x + 3y = 15$

Write an equation for each line being described. Write your answer in slope-intercept form.

25. Through point $(3, 4)$,
parallel to $y = -x + 2$

Through point $(4, 1)$,
perpendicular to $y = -6x + 2$

26. Through point $(-2, 2)$,
perpendicular to $y = 4$

Through point $(-5, -1)$,
parallel to $y = \frac{3}{5}x - 4$

27. Through point $(-3, 2)$,
parallel to $5x + 4y = -4$

Through point $(-1, -1)$,
perpendicular to $x + 3y = 3$

28. Through point $(4, 4)$,
perpendicular to $4x + 7y = 21$

Through point $(1, 1)$,
parallel to $-x + y = -4$

For each pair of points below, find the midpoint.

29. $(1, -2)$ and $(1, -6)$

$(-9, 2)$ and $(-3, 2)$

30. $(10, 9)$ and $(-10, 4)$

$(-8, 3)$ and $(2, 2)$

31. $(7, 3)$ and $(8, 3)$

$(9, 7)$ and $(1, -3)$

Find the distance between each pair of points. Write radicals in simplified form.

32. $(7, -2)$ and $(-1, -8)$

$(-3, -4)$ and $(-7, -8)$

33. $(1, -2)$ and $(7, -4)$

$(-4, -1)$ and $(-8, -4)$

34. $(-6, -5)$ and $(3, 4)$

$(4, 0)$ and $(-4, 4)$

35. Find the lengths of the sides of figure *PQRS* at the right.

side *PQ* _____

side *QR* _____

side *RS* _____

side *SP* _____

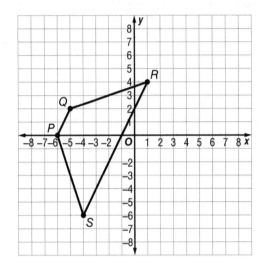

36. A trapezoid is a figure with exactly two parallel sides. Is figure *PQRS* in problem 35 a trapezoid? Explain.

37. Draw two line segments connecting points *A* and *B* and points *C* and *D*. What is the distance between the midpoints of *AB* and *CD*?

midpoint of *AB* _____

midpoint of *CD* _____

distance between midpoints:

_____

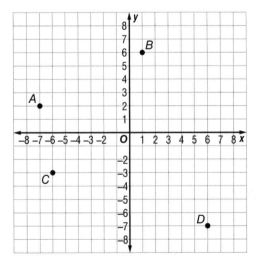

PROBLEM SOLVING WITH MULTIPLE VARIABLES

In the previous section, you saw how the solution to an equation with two variables, x and y, can be represented geometrically by a line on the coordinate plane. Now you will learn how to harness the power of equations with two variables to represent and solve a variety of applications.

Graphing a System of Equations

In this lesson, you will learn to

- Use graphing to find or estimate a solution to a system of equations.

Often, more than one equation can be written about the same situation. Consider these statements about two variables x and y: A number y is three times the value of x, and the sum of x and y is 8.

Both of these statements can be written with symbols.

y is three times the value of x

$$y = 3x$$

the sum of x and y is 8

$$x + y = 8$$

These two equations are connected because the variables x and y represent the same values in both equations. Two or more equations that are connected form a **system of equations.** A solution to a system of equations is the ordered pair (x, y) that satisfies both equations.

Note: In the system of equations at the right, the bracket shows that the equations form a system. A bracket is not needed if the problem states that the equations are a system.

$$\begin{cases} y = 3x \\ x + y = 8 \end{cases}$$

You have probably noticed that both of these equations are exactly like the kind you graphed in the previous section. You can use graphing to find the solution to this problem. First, write each equation in terms of y so that it begins with $y =$. Then graph the equations on the same coordinate plane.

Graph $y = 3x$.

Graph $x + y = 8$.

Find the point where the lines intersect.

The point **(2, 6)** is the solution to the system. This point makes both equations true.

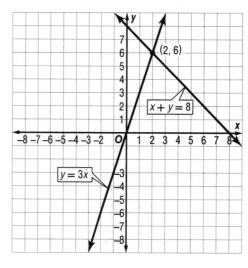

$y = 3x$	$x + y = 8$
$6 = 3 \cdot 2$	$2 + 6 = 8$
$6 = 6$ True	$8 = 8$ True

The check step shows that you can solve a system of equations by graphing both equations and finding the point of intersection. Graphing is a convenient method for solving a system when the values for x and y are integers.

Choose the method of graphing that seems the most convenient for the situation. You can use any of the graphing methods in the last section. The graphing steps will not be shown in detail in this section.

EXAMPLE 1 Solve the system by graphing: $x + y = -1$
$$3x - 2y = -8$$

STEP 1 Graph both equations.

STEP 2 Find the point of intersection. The solution to the system is **(−2, 1)**.

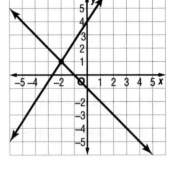

CHECK Substitute the values for x and y into <u>both</u> equations.

$$x + y = -1$$
$$-2 + 1 = -1$$
$$-1 = -1 \text{ True}$$

$$3x - 2y = -8$$
$$3(-2) - 2(1) = -8$$
$$-6 - 2 = -8$$
$$-8 = -8 \text{ True}$$

Solve each system of equations by graphing.

1. $4x - y = 7$
   $x + y = -2$

   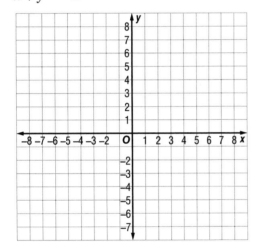

   $4x - 5y = 15$
   $x + y = 6$

   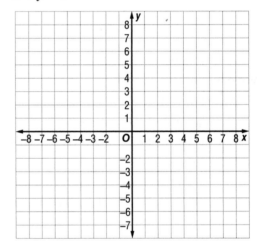

2. $9x + 8y = -24$
   $x - 4y = -32$

   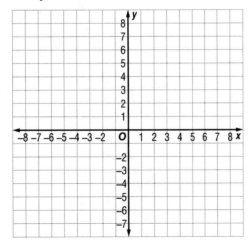

   $11x - 4y = 36$
   $5x + 4y = 28$

   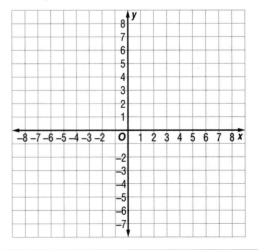

3. $3x + y = -1$
   $2x - y = -4$

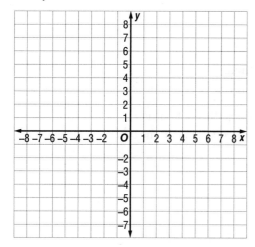

$x - y = 3$
$x = -3$

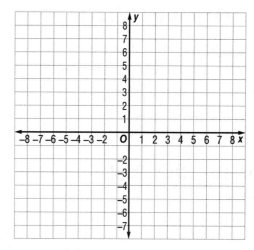

4. $x + 3y = -6$
   $5x - 3y = 24$

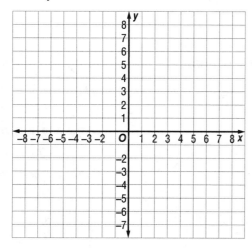

$x - 2y = 8$
$9x - 2y = -8$

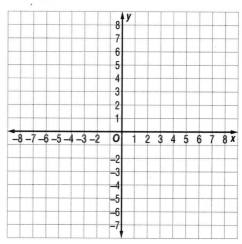

5. $x - 4y = 24$
   $x + y = -1$

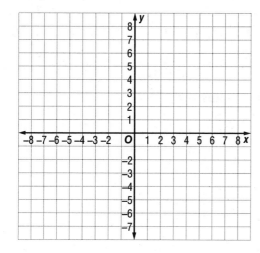

$y = -4$
$5x - y = -6$

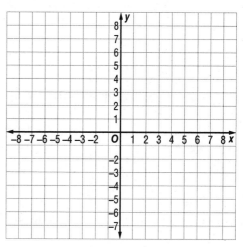

Solving a System of Equations Using Algebra

In this lesson, you will learn to

- Solve a system of equations using the substitution method.
- Solve a system of equations using the elimination method.

There are two algebraic methods for solving systems of equations. The first one is called **substitution** because you substitute an expression from one equation into the other.

Substitution Method

EXAMPLE 1   Solve the system by substitution: $-5x - 2y = 8$
$y = x + 10$

Look at the second equation in the system, $y = x + 10$. This equation tells you that the value of y is $x + 10$. When one equation is solved for a single variable, you can take that value and substitute it into the other equation.

STEP 1   Solve one equation for a single variable. The second equation is already solved for y: $y = x + 10$.

STEP 2   Substitute the expression into the remaining equation. Write $x + 10$ in place of y.

STEP 3   Solve the new equation for the variable.

$$-5x - 2y = 8$$
$$-5x - 2(x + 10) = 8$$
$$-5x - 2x - 20 = 8$$
$$-7x - 20 = 8$$
$$-7x = 28$$
$$x = -4$$

STEP 4   Go back to either of the original equations and substitute -4 for x.

$$y = x + 10$$
$$y = -4 + 10$$
$$y = 6$$

Write the values of x and y as an ordered pair. The solution is **(−4, 6)**.

CHECK   Remember to check the solution in *both* equations.

$$-5x - 2y = 8$$
$$-5(-4) - 2(6) = 8$$
$$20 - 12 = 8$$
$$8 = 8 \text{ True}$$

$$y = x + 10$$
$$6 = -4 + 10$$
$$6 = 6 \text{ True}$$

Solve each system using the substitution method. Check your answer.

1. $y = -5x + 15$
   $-2x - 6y = -6$

   $y = 6x + 12$
   $3x + 2y = 9$

   $x = -5$
   $x + 3y = 10$

2. $-2x - y = -5$
   $y = -6x + 17$

   $x = -3y + 5$
   $-6x - y = 4$

   $y = 5x + 15$
   $-4x + 3y = 1$

Problem solving is about making good decisions. In this next example, you need to decide which equation to write in terms of a single variable.

EXAMPLE 2 Solve the system by substitution: $6x - 5y = -20$
$x - 7y = 9$

You need to write one of these equations in terms of either x or y. Each term in the equation $6x - 5y = -20$ has a number coefficient that would make isolating the variable more difficult. It would be easier to solve the second equation $x - 7y = 9$ in terms of x.

STEP 1 Solve one equation for a single variable.

$$x - 7y = 9$$
$$x = 7y + 9$$

STEP 2 Substitute the expression into the remaining equation.

STEP 3 Solve the new equation for the variable.

$$6x - 5y = -20$$
$$6(7y + 9) - 5y = -20$$
$$42y + 54 - 5y = -20$$
$$37y = -74$$
$$y = -2$$

STEP 4 Go back to either of the original equations and substitute -2 for y. This example uses the second equation because it looks easier to work with.

$$x - 7y = 9$$
$$x - 7(-2) = 9$$
$$x + 14 = 9$$
$$x = -5$$

Write the values of x and y as an ordered pair. The solution is **(−5, −2)**.

Check the solution in both equations:

$$6x - 5y = -20 \qquad\qquad x - 7y = 9$$
$$6(-5) - 5(-2) = -20 \qquad -5 - 7(-2) = 9$$
$$-30 + 10 = -20 \qquad\qquad -5 + 14 = 9$$
$$-20 = -20 \text{ True} \qquad\qquad 9 = 9 \text{ True}$$

Is it really important to check the solution in both equations? Yes. Remember that each linear equation has infinitely many solutions. You are looking for the point where the graphs of the lines intersect—the one solution that satisfies both equations. If your solution only works in one of the equations, then it is not the point of intersection of the graphs, and it is not a solution to the system.

Solve each system using the substitution method. Check your answer.

3. $x + 4y = 27$ 　　　　　$5x - 3y = -2$ 　　　　　$x - 3y = 1$
　　$-5x + 7y = 0$ 　　　　$2x + y = -3$ 　　　　　$-3x + 2y = 4$

4. $x + 2y = 28$ 　　　　　$-4x + y = 16$ 　　　　　$-x + y = -10$
　　$-10x + 8y = 0$ 　　　　$-2x - 7y = 8$ 　　　　　$-5x - 3y = -18$

5. $6x - 4y = -26$ 　　　　$-4x - 7y = 21$ 　　　　　$-5x - 6y = 21$
　　$x + y = 9$ 　　　　　　$x - y = 14$ 　　　　　　$2x + y = 0$

6. $-8x - 7y = 16$ $x - 5y = -25$ $8x + 7y = 11$
    $x + 4y = -27$ $-3x - 5y = 15$ $x - 6y = 22$

7. $5x - 3y = -8$ $x - 7y = -5$ $2x - 5y = 15$
    $x + 2y = -12$ $5x - 3y = 7$ $x + 5y = -15$

Depending on your choices, you may sometimes encounter fractions in an equation. Don't worry. This does not mean you have made a mistake. You can always multiply both sides of the equation by the lowest common denominator to clear the fractions.

EXAMPLE 3   Solve the system by substitution: $5x - 6y = 25$
$-3x + 5y = -29$

STEP 1   Write the first equation in terms of x. (Remember, you can solve either equation in terms of x or y. The final answer will be the same.)

$$5x - 6y = 25$$
$$5x = 6y + 25$$
$$\frac{5x}{5} = \frac{6y}{5} + \frac{25}{5}$$
$$x = \frac{6}{5}y + 5$$

STEP 2   Substitute the expression into the remaining equation.

$$-3\left(\frac{6}{5}y + 5\right) + 5y = -29$$

STEP 3   Solve the new equation for the variable.

$$-\frac{18}{5}y - 15 + 5y = -29$$

Multiply both sides of the equation by 5 to remove the fraction. Notice that every term must be multiplied by 5.

$$5\left(-\frac{18}{5}y - 15 + 5y\right) = 5(-29)$$
$$-18y - 75 + 25y = -145$$
$$7y = -70$$
$$y = -10$$

STEP 4   Go back to either of the original equations and substitute -10 for y.

$$5x - 6(-10) = 25$$
$$5x + 60 = 25$$
$$5x = -35$$
$$x = -7$$

The solution is **(−7, −10)**. You can check this answer on your own.

Remember that the solution to a system can be found by graphing both equations. The point of intersection represents the solution, or ordered pair, that satisfies both equations. With that in mind, what would it mean if two equations failed to intersect?

EXAMPLE 4   Solve the system by substitution: $x + 2y = 10$
$x + 2y = -6$

What do you notice right away? The left sides of both equations are identical.

Rewrite both equations in slope-intercept form.

$x + 2y = 10$ is rewritten as $y = -\frac{1}{2}x + 5$.

$x + 2y = -6$ is rewritten as $y = -\frac{1}{2}x - 3$.

Both equations have the same slope. The equations represent parallel lines. Looking at the graph of the equations, you can see that the lines will never intersect.

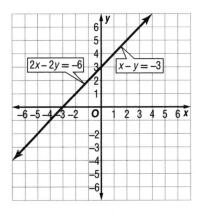

This system has **no solution.**

A system that has no solution is an **inconsistent system.** A system that has one solution is a **consistent system.**

If you attempt to solve an inconsistent system algebraically, the variables will be eliminated and you will get a false statement. Try following the usual steps to solve Example 4.

STEP 1 Rewrite one equation in terms of x.

STEP 2 Substitute this expression for x in the remaining equation.

The system has **no solution** and is **inconsistent.**

$$x + 2y = 10$$
$$x = -2y + 10$$
$$-2y + 10 + 2y = -6$$
$$10 = -6 \text{ False}$$

There is one additional kind of system. A **dependent system** has infinitely many solutions.

EXAMPLE 5 Solve the system by substitution: $x - y = -3$
$$2x - 2y = -6$$

STEP 1 Rewrite the first equation to solve for x.

STEP 2 Substitute this expression for x in the second equation.

$$x - y = -3$$
$$x = y - 3$$
$$2(y - 3) - 2y = -6$$
$$2y - 6 - 2y = -6$$
$$-6 = -6 \text{ True}$$

The statement is true, but the variables have both vanished. This system is **dependent. Every ordered pair that satisfies one equation also satisfies the other.** A graph of this system looks like a single line because the lines coincide. They occupy the same space.

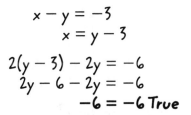

Solve each system using the substitution method. Write *inconsistent* or *dependent* if there is not a single solution to the system.

8. $x + 4y = 27$ $3x + 6y = 0$ $-10x + 2y = 14$
    $2x + 8y = 0$ $-3x - 9y = 9$ $2x - 5y = 11$

9. $2x - 3y = 0$ $-4x + y = 16$ $4x + 5y = -26$
    $-5x + 3y = 0$ $-2x - 7y = 8$ $-4x - 9y = 18$

10. $5x - 2y = -21$
$\quad\ -4x + 7y = 6$

$2x + 2y = 0$
$-3x - y = -4$

$x - 3y = 1$
$-3x + 9y = -3$

11. $8x - 12y = 16$
$\quad\ -2x + 3y = -4$

$7x + 5y = 6$
$-x + 4y = 18$

$x + y = -10$
$-x - y = 15$

12. $x + 2y = 6$
$\quad\ 5x + 10y = 0$

$-4x - 6y = 16$
$-2x - 3y = 8$

$2x - 9y = -2$
$-4x - 6y = 4$

Elimination Method

Another method for solving systems of equations is **elimination.** In this method, the goal is to eliminate either variable x or y by adding.

EXAMPLE 6 Solve the system by elimination: $5x - 9y = -13$
$\qquad\qquad\qquad\qquad\qquad\qquad\qquad -3x + 9y = 15$

What would happen if you simply added these equations?

The variable y is eliminated because $-9y + 9y = 0$. These terms are **additive inverses**—opposites whose sum is 0.

$$\begin{array}{r} 5x - 9y = -13 \\ -3x + 9y = \ \ \ 15 \\ \hline 2x \qquad\quad = \ \ \ 2 \end{array}$$

You can now solve $2x = 2$ for the value of x.

$$2x = 2$$
$$x = 1$$

Then substitute the value of x into one of the original equations. Substitution into the first equation is shown here.

The solution to the system is **(1, 2).** You can check your answer in both equations just as you did with the substitution method.

$$5x - 9y = -13$$
$$5(1) - 9y = -13$$
$$5 - 9y = -13$$
$$-9y = -18$$
$$y = 2$$

Why does the elimination method work? The **additive property of equality** states that you can add the same quantity to both sides of an equation. An equation is a statement that two expressions are equal. This means that both sides of an equation are equal in value. So when you add the sides of one equation to a second equation, you are adding the same quantity to both sides of that second equation.

Many systems do not start out with additive inverses. You can, however, perform multiplication to one or both of the equations to create additive inverses. Then you can solve the system by elimination. Often, you can create additive inverses by multiplying every term in one of the equations by −1.

EXAMPLE 7 Solve the system by elimination: $-4x + 5y = 0$
$\qquad\qquad\qquad\qquad\qquad\qquad\qquad -6x + 5y = -10$

There is a $+5y$ term in both equations. If one were a $-5y$ instead, there would be additive inverses. To change one $+5y$ into a $-5y$, multiply both sides of the second equation by −1. This has the effect of changing every sign in the second equation.

$$-6x + 5y = -10$$
$$(-1)(-6x + 5y) = (-1)(-10)$$
$$6x - 5y = 10$$

You can now rewrite the system and solve.

$$-4x + 5y = 0$$
$$\underline{6x - 5y = 10}$$
$$2x \quad = 10$$
$$x = 5$$

Substitute 5 for x in one of the original equations.

Note: It is very important to substitute your answer into one of the original equations in the system. If you substitute a value into an equation you have changed, you may introduce an error into your work.

$$-4x + 5y = 0$$
$$-4(5) + 5y = 0$$
$$-20 + 5y = 0$$
$$5y = 20$$
$$y = 4$$

The solution to the system is **(5, 4)**.

Solve each system using the elimination method.

13. $4x - y = -1$
$\quad 6x + y = 1$

$-5x - 2y = 17$
$5x + 5y = -20$

$7x + 4y = 13$
$-7x + 3y = 22$

14. $x - 5y = -4$
$\quad -2x - 5y = -22$

$9x + 8y = 25$
$9x + 10y = 11$

$x + 2y = 17$
$x + y = 8$

15. $-5x - 3y = -23$
$\quad x + 3y = -5$

$-10x + 4y = 10$
$-10x - 4y = 10$

$3x + 3y = 21$
$3x + y = 13$

16. $-4x - 6y = -20$
$\quad -4x + 4y = 20$

$-x + y = 1$
$-5x + y = -7$

$8x + 5y = 18$
$8x - 9y = -10$

You can always use multiplication to rewrite a system so that the elimination method will work. Remember, when you multiply both sides of an equation by the same quantity, the equation looks different, but its value remains the same.

EXAMPLE 8 Solve the system by elimination: $\quad 3x + 7y = -1$
$\quad\quad\quad\quad\quad\quad\quad\quad\quad\quad\quad\quad -x - 10y = 8$

STEP 1 Choose a variable to eliminate. Multiply to create additive inverses. You can eliminate the variable x if you multiply the second equation by 3.

$$(3)(-x - 10y) = (3)(8)$$
$$-3x - 30y = 24$$

STEP 2 Rewrite the system and solve for the remaining variable.

$$3x + 7y = -1$$
$$\underline{-3x - 30y = 24}$$
$$-23y = 23$$
$$y = -1$$

STEP 3 Substitute the value into one of the original equations and solve.

$$3x + 7y = -1$$
$$3x + 7(-1) = -1$$
$$3x - 7 = -1$$
$$3x = 6$$
$$x = 2$$

The solution is **(2, −1)**.

You may need to perform multiplication on both equations to create additive inverses and use the elimination method. Multiply carefully and remember to use one of the original equations to find the value of the remaining variable.

EXAMPLE 9   Solve the system by elimination: $-5x + 7y = -10$
$2x - 6y = 4$

STEP 1   Choose a variable to eliminate. To create additive inverses for the variable x, multiply the first equation by 2 and the second equation by 5.

Equation 1 $-5x + 7y = -10$
$(2)(-5x + 7y) = (2)(-10)$
$-10x + 14y = -20$

Equation 2 $2x - 6y = 4$
$(5)(2x - 6y) = (5)(4)$
$10x - 30y = 20$

STEP 2   Rewrite the system and solve for the remaining variable.

$$-10x + 14y = -20$$
$$10x - 30y = 20$$
$$\overline{-16y = 0}$$
$$y = 0$$

STEP 3   Substitute the value into one of the original equations and solve.

$$-5x + 7y = -10$$
$$-5x + 7(0) = -10$$
$$-5x = -10$$
$$x = 2$$

The solution is **(2, 0)**.

You now have three methods for solving systems of equations: graphing, substitution, and elimination. They all work. With practice, you will be able to choose the best one for a given situation.

Remember that not all systems have one solution. An inconsistent system has no solution. A dependent system has infinitely many solutions.

Solve each system using the method of your choice. Write *inconsistent* or *dependent* if the system does not have a single solution.

17. $2x + 5y = 25$ $-7x + 12y = 15$ $-x + 5y = -1$
$8x - 6y = 22$ $4x - 4y = -20$ $-5x + 4y = -5$

18. $-x + 2y = 15$ $7x + 2y = -18$ $-2x + 2y = -14$
$-6x - 3y = 0$ $14x - y = 9$ $x + y = 5$

19. $-2x + 4y = -1$ $-4x - y = -12$ $-7x + 4y = -11$
$x - 2y = -7$ $-7x + y = 1$ $x - 3y = -13$

20. $6x - y = 0$ $2x + 7y = -9$ $3x - 8y = -29$
$-6x + y = 0$ $3x - 2y = 24$ $-2x + 6y = 24$

21. $-4x - 7y = 0$ $-2x - 5y = -2$ $3x - 3y = -9$
     $-7x - 5y = 0$ $3x + 8y = 4$ $-5x + 5y = 15$

22. $-7x - 4y = -3$ $5x - y = 14$ $2x - y = -11$
     $4x + 5y = -1$ $6x - 5y = -6$ $3x - y = -15$

23. $7x - 8y = 19$ $9x - 21y = 7$ $-8x - y = -22$
     $-2x - 4y = -18$ $3x - 7y = 5$ $8x + y = 22$

24. $3x + 6y = 0$ $6x + 5y = 10$ $5x - 2y = -11$
     $5x - y = 11$ $-x - 2y = -11$ $3x - 4y = -1$

Synthesis   The same basic strategies can be employed to solve a system of three equations with three variables. Study the example to see how it is done. The equations in the example are labeled A, B, and C.

EXAMPLE 10   Solve the system: $x - 3y + 2z = 1$ (A)
                                           $-x + y + 3z = 7$ (B)
                                           $-x - 5y - z = -7$ (C)

STEP 1 Combine A and B to eliminate x, and form equation D.

STEP 2 Combine A and C to eliminate x, and form equation E.

STEP 3 Using equations D and E, solve for y and z as you would solve any 2-variable system.

$$\begin{array}{ll} x - 3y + 2z = 1 & (A) \\ -x + y + 3z = 7 & (B) \\ \hline -2y + 5z = 8 & (D) \end{array}$$

$$\begin{array}{ll} x - 3y + 2z = 1 & (A) \\ -x - 5y - z = -7 & (C) \\ \hline -8y + z = -6 & (E) \end{array}$$

$$\begin{array}{ll} -2y + 5z = 8 & (D) \\ -8y + z = -6 & (E) \end{array}$$

Complete Step 3 on your own. You should find that **$y = 1$** and **$z = 2$**. Now go back to one of the original equations and substitute the values of y and z to solve for x.

$$\begin{array}{l} x - 3y + 2z = 1 \\ x - 3(1) + 2(2) = 1 \\ x - 3 + 4 = 1 \\ x + 1 = 1 \\ \boldsymbol{x = 0} \end{array}$$

The solution set (x, y, z) to the system is **(0, 1, 2)**.

Solve each system.

25. $-3x + 2y - 2z = -6$ $-2x + 3y - 6z = 14$
        $3x + 4y - z = 15$ $-5x - 6y + 2z = 8$
        $3x + 5y + 3z = 3$ $2x - y + 4z = -10$

Solving Word Problems Using Systems

In this lesson, you will learn to

- Solve a word problem using a system of equations.

Word problems can often be solved using a system of equations. To solve a problem using a system, you must be able to write the same number of equations as you have variables. In other words, if you want to use two variables, you need to be able to write two equations.

The 5-step problem-solving process on page 42 will also work when you solve problems using a system. The five steps are listed here with a few small changes.

Problem-Solving Plan

STEP 1 Read and organize the information in the problem.

STEP 2 Assign variables and write expressions.

STEP 3 Write related equations to form a system.

STEP 4 Solve the system and check.

STEP 5 Apply the solution to answer the question stated in the problem.

EXAMPLE 1 A softball team played 60 games. The team won 24 more games than it lost. How many games did the team lose?

STEP 1 This problem includes three quantities: wins, losses, and total games played.

STEP 2 Let W represent the number of wins, L represent the number of losses, and $W + L$ represent the total games played.

STEP 3 You know that the total games played is equal to 60 and that the difference between the wins and the losses is 24.

Write a system:
$$W + L = 60$$
$$W - L = 24$$

STEP 4 Solve.

Use elimination:
$$\begin{aligned} W + L &= 60 \\ W - L &= 24 \\ \hline 2W &= 84 \\ W &= 42 \end{aligned}$$

Solve for L:
$$\begin{aligned} W + L &= 60 \\ 42 + L &= 60 \\ L &= 18 \end{aligned}$$

Check:
$$42 + 18 = 60$$
$$60 = 60$$
$$42 - 18 = 24$$
$$24 = 24$$

STEP 5 Apply the solution. The question asks: How many games did the team lose? The answer is **18 games.**

How is this different from solving problems with one variable? When you work with one variable, you must express every fact in the problem in terms of that one variable. When you work with a system, you can use different variables for different facts. There are many approaches to problem solving. Whether you use one variable or write a system is up to you. There isn't one right way to solve a problem. Use the way that makes the most sense to you.

Solve. A system is provided for selected problems.

1. The length of a rectangle is 2 inches longer than its width. The sum of the length and the width is 14 inches. What is the width of the rectangle?

$$\begin{cases} L + W = 14 \\ L - W = 2 \end{cases}$$

2. Phillip has 130 coins, all dimes and nickels. He has 40 more dimes than nickels. How many dimes does Phillip have?

3. Isabel is 3 years older than Hal. The sum of their ages is 17. How old is Hal?

4. One number (x) is 6 more than twice another (y). If the sum of the numbers is 51, what are the two numbers?

$$\begin{cases} x = 6 + 2y \\ x + y = 51 \end{cases}$$

5. A parking lot has 66 parking spaces. If there are 16 more compact spaces than regular spaces, how many spaces are there for compact cars?

6. In a basketball game, the winning team scored 18 more points than the losing team. If there were 188 points scored in the game, how many points did the losing team score?

7. Tickets to an awards banquet are $25 for parents and $15 for students. Ticket sales for the banquet totaled $2,225. There were 25 fewer student tickets sold than parent tickets. How many student tickets were sold?

$$\begin{cases} 15s + 25p = 2,225 \\ s = p - 25 \end{cases}$$

8. The combined weight of 4 red bricks and 5 gray bricks is 44 pounds. The combined weight of 6 red bricks and 3 gray bricks is 48 pounds. How much do 5 red bricks weigh?

Using systems of equations is a convenient way to solve digit problems. As you know, the value of a digit depends on its placement in a number. Consider the digit 5 in the numbers 53 and 35. In 53, the value of the digit 5 is $5 \cdot 10 = 50$, while its value is only $5 \cdot 1 = 5$ in the number 35.

Digit problems often compare the values of numbers before and after digits trade places. Study the next example to see how to handle digit problems.

EXAMPLE 2 The sum of the digits of a two-digit number is 14. If the digits are reversed, the new number is 18 greater than the original number. What is the original number?

STEP 1 There are two digits in both numbers: the tens digit and the ones digit. You need to remember that the original number is the smaller one.

STEP 2 Use variables that are easy to remember and will not confuse you. For example, it makes sense to use T for tens and O for ones. You could, however, easily confuse the letter O with the number 0, so instead use U for the ones place.

Write expressions for the new number and the original number by attaching value to the digits. The original number can be written $10T + U$. In the new number, the digits are reversed. The new number can be written $10U + T$.

STEP 3 You know the sum of the digits is 14. You also know that the new number is 18 greater than the original. Use the expressions from Step 2 to write a system.

$$T + U = 14$$
$$(10U + T) - (10T + U) = 18$$

STEP 4 Solve.

First, simplify the second equation:

$$(10U + T) - (10T + U) = 18$$
$$10U + T - 10T - U = 18$$
$$-9T + 9U = 18$$

Rewrite the system:

$$T + U = 14$$
$$-9T + 9U = 18$$

Rewrite the first equation so that you can use substitution:

$T + U = 14$ is rewritten as
$$T = 14 - U$$

Solve by substituting $14 - U$ for T in the second equation:

Substitute 8 for U in the simpler equation. $T + 8 = 14$, so $T = 6$.

$$-9(14 - U) + 9U = 18$$
$$-126 + 9U + 9U = 18$$
$$18U = 144$$
$$U = 8$$

STEP 5 Apply the solution to answer the question. **The original number (the smaller one) is 68.** To check your answer, make sure that 86 is 18 greater than 68.

Solve. A system is provided for selected problems.

9. When you reverse the digits in a two-digit number, the value of the new number is 27 greater than that of the original number. Find the number if the sum of its digits is 13.

$$\begin{cases} T + U = 13 \\ (10U + T) - (10T + U) = 27 \end{cases}$$

10. The sum of the digits of a certain two-digit number is 3. When you reverse its digits, you decrease the number by 9. What is this number?

(**Hint:** The phrase *certain number* refers to the original number.)

11. When you reverse the digits of a certain two-digit number, you increase its value by 36. What is this number if the sum of its digits is 6?

12. The sum of the digits of a certain two-digit number is 11. If you reverse its digits, the new number is 27 greater than the original number. Find the new number.

13. When you reverse the digits of a certain two-digit number, you decrease its value by 54. What is the number if the sum of its digits is 12?

14. The sum of the digits of Frank's age is 9. Frank noticed that when he reverses the digits of his age, he gets his grandfather's age. Frank's grandfather is 63 years older than he is. How old is Frank?

Another type of problem asks about the numbers in a fraction. Let n represent the numerator and d represent the denominator. Let $\frac{n}{d}$ represent the fraction. Steps 1 and 2 in problem 15 are completed for you, and Step 3 is started for you.

15. The denominator of a fraction is 5 more than its numerator. If you add three to both the numerator and the denominator, the value of the resulting fraction equals $\frac{1}{2}$. What is the original fraction?

STEP 1 Write the system:

$$d - n = 5$$
$$\frac{n + 3}{d + 3} = \frac{1}{2}$$

STEP 2 Cross multiply to simplify the second equation:

$$2(n + 3) = 1(d + 3)$$
$$2n + 6 = d + 3$$
$$-d + 2n = -3$$

STEP 3 Write the new system and solve:

$$d - n = 5$$
$$-d + 2n = -3$$

16. The denominator of a fraction is 4 more than its numerator. If you increase the denominator by 7 and decrease the numerator by 7, the resulting fraction has a value of $\frac{1}{4}$. What is the original fraction?

17. The sum of the numerator and the denominator of a fraction is 20. If you subtract 2 from the numerator and add 2 to the denominator, the value of the resulting fraction is $\frac{1}{3}$. What is the original fraction?

Solving Rate Problems Using Systems

In this lesson, you will learn to

- Solve word problems involving ratios.
- Understand the difference between direct and indirect variation.

You already know quite a bit about ratio and proportion. You know that a ratio is usually written as a fraction. You know that a proportion is a statement that two ratios are equal. You also know that you can use cross multiplication to solve a proportion.

EXAMPLE 1 The ratio of men to women in a business class is 3 to 5. If there are 48 students in the class, how many are women?

This is a fairly straightforward problem. Let m represent men and w represent women.

STEP 1 Write two equations because there are two variables. The first equation is the ratio given in the problem, $\frac{m}{w} = \frac{3}{5}$. Use cross multiplication and the properties of equality to write it in standard form.

$$\frac{m}{w} = \frac{3}{5}$$
$$5m = 3w$$
$$5m - 3w = 0$$

The second equation uses the fact that there is a total of 48 students in the class.

$$m + w = 48$$

Write the system.

$$\begin{cases} 5m - 3w = 0 \\ m + w = 48 \end{cases}$$

STEP 2 Solve the system.

Rewrite $m + w = 48$ as $m = 48 - w$, and then use substitution.

$$5m - 3w = 0$$
$$5(48 - w) - 3w = 0$$
$$240 - 5w - 3w = 0$$
$$240 = 8w$$
$$w = 30$$

There are **30 women** in the class.

To check this answer, first find the number of men: $48 - 30 = 18$. The ratio $\frac{18}{30}$ is equal to $\frac{3}{5}$, so the answer checks.

Try solving a few ratio problems using a system of equations. Remember, you can choose any method you like to solve the system. Both substitution and elimination will work.

Solve.

1. The ratio of the width to the length of a rectangle is 2 to 3. If you subtract 3 inches from the width and add 3 inches to the length, the new ratio of width to length will be 1 to 4. What is the length of the original rectangle?

2. A cashier has dimes and nickels worth $7.80. If the ratio of dimes to nickels is 4 to 5, how many nickels does the cashier have?

3. A bag contains red and blue marbles. There are 2 blue marbles for every 9 red ones. If you put 10 blue marbles in the bag and you take out 15 red ones, there will be 1 blue marble for every 3 red ones. How many blue marbles were originally in the bag?

4. At a restaurant, the ratio of lunch customers to dinner customers is 3 to 4. If the total number of customers in a day is 525, how many customers came during lunch?

5. A college class is comprised of only juniors and seniors. The ratio of juniors to seniors is 1 to 2. If 2 of the juniors drop the class and 1 senior adds the class, the ratio of juniors to seniors changes to 2 to 5. How many students were originally in the class?

6. At a sporting event, the ratio of peanuts sold to nachos sold was 3 to 2. If 50 customers who bought peanuts had instead bought nachos, then the ratio would have been 1 to 1. How many customers bought nachos?

To work with ratios and proportions algebraically, you need to understand the concept of variation. Consider the two situations below.

| Direct Variation | Suppose you take a part-time job where you earn an hourly wage. If you work more hours, you can expect to earn more money. | Hours Worked   Total Pay |
| Inverse (or Indirect) Variation | Suppose you start a project that should take 10 hours. If your friends help, the amount of time the project takes should decrease. | Number of Workers   Hours the Job Takes |

Both of these examples have a constant (k). A constant is an unchanging value. In the example of direct variation, the constant is the hourly wage. In the example of inverse variation, the constant is the job itself. The project doesn't change even if the number of workers changes.

In algebra, the variable k is used to represent a constant. Direct and inverse variation are defined algebraically using x, y, and the constant k.

| Direct Variation: $y = kx$ |
| *y varies directly as x.* |

| Inverse Variation: $y = \frac{k}{x}$ |
| *y varies inversely as x.* |

Try putting these ideas to work.

EXAMPLE 2 Write an equation of variation where y varies directly as x, and $y = 35$ when $x = 7$.

STEP 1 This is a direct variation problem. Substitute the given values to find the constant k.

$$y = kx$$

$$35 = k \cdot 7$$

STEP 2 Substitute 5 for k in the direct variation equation: $y = 5x$.

$$k = \frac{35}{7}$$

This equation of variation is an example of a function, which means it forms the rule that governs how the variables in this situation interact. You can think of it as a formula that you have created. You will learn about functions in more detail on pages 204–210.

$$k = 5$$

EXAMPLE 3 Write an equation of variation where y varies inversely as x, and $y = 6$ when $x = 9$.

$$y = \frac{k}{x}$$

STEP 1 This is an inverse variation problem. Substitute the given values to find the constant k.

$$6 = \frac{k}{9}$$

STEP 2 Substitute 54 for k in the inverse variation equation: $y = \frac{54}{x}$.

$$k = 54$$

Write the equations of variation as directed. If necessary, express k as a decimal.

7. y varies directly as x,
   and $y = 200$ when $x = 25$

   y varies directly as x,
   and $y = 10$ when $x = 40$

8. y varies inversely as x,
   and $y = 3$ when $x = 15$

   y varies inversely as x,
   and $y = 16$ when $x = 2$

9. y varies directly as x,
   and $y = 50$ when $x = 20$

   y varies inversely as x,
   and $y = 4$ when $x = 7$

10. y varies inversely as x,
    and $y = 8$ when $x = 100$

    y varies directly as x,
    and $y = 200$ when $x = 10$

11. y varies directly as x,
    and $y = 8$ when $x = 0.5$

    y varies directly as x,
    and $y = 64$ when $x = 8$

12. y varies inversely as x,
    and $y = 12$ when $x = 3$

    y varies inversely as x,
    and $y = 48$ when $x = 6$

Solving Mixture and Work Problems

In this lesson, you will learn to

- Solve rate applications involving mixtures and solutions.
- Solve rate applications known as work problems.

Have you ever used a recipe to cook or bake? If so, you were working with rates. If you used a cake recipe, the recipe told how much of each ingredient you need for one cake.

What does this have to do with rates? The numbers in the recipe are constants (k). The amount of an ingredient that you need (y) varies directly as the number of cakes you want to bake (x). If you double the ingredients, you double the number of cakes.

Mixtures and solutions are related to recipes. When you combine dry ingredients, you are making a mixture. When you combine liquid ingredients, you are making a solution.

EXAMPLE 1 Peanuts sell for $2 per pound and cashews sell for $7 per pound. A company wants to make 100 pounds of a mix that would be valued at $5 per pound. How many pounds of each kind of nut should go in the mix?

STEP 1 Let p represent the pounds of peanuts and c represent the pounds of cashews.

You can make a chart to represent the information in the problem.

	pounds	× price	= value
Peanuts	p	2	$2p$
Cashews	c	7	$7c$
Mixture	100	5	500

STEP 2 Write two equations. The first equation comes from the first column; it sets p and c equal to the total pounds in the mix.

The second equation uses the total value of the ingredients. It comes from the third column.

$$\begin{cases} p + c = 100 \\ 2p + 7c = 500 \end{cases}$$

STEP 3 Solve the system. The elimination method is shown. Multiply each term in the first equation by −2 so that you can eliminate the variable p.

$$\begin{array}{r} -2p + -2c = -200 \\ 2p + 7c = 500 \\ \hline 5c = 300 \\ c = 60 \end{array}$$

After solving for c, return to the original first equation and solve for p.

The company should use **40 pounds of peanuts** and **60 pounds of cashews** to make a 100-pound mixture valued at $5 per pound.

$$p + 60 = 100$$
$$p = 40$$

Do you see how this problem connects to direct variation? The price is the constant k, which is given in the problem. There is a direct relationship between the facts in the problem. If the number of pounds goes up, then the value goes up. Compare $y = kx$ to $value = price \times pounds$.

Percents are generally used when dealing with solutions. Remember to change percents to decimals when you write equations.

EXAMPLE 2 Shayna wants to make a 50% acid solution. She has already poured 8 mL of a 35% acid solution into a beaker. How many milliliters of a 60% acid solution must she add to the beaker to create the desired mixture?

STEP 1 Use a chart to organize the information. Notice that the columns of this chart are similar to those in Example 1. Because you are given more information in this problem, you need only one variable. The given information from the problem has been entered on the chart.

	total mL ×	% saline =	mL of acid
1st solution	8	35%	
2nd solution	x	60%	
Total mixture		50%	

Now complete the chart. Add down to fill in the bottom row. Multiply across to fill in the third column. Here is the completed chart. Make sure you understand how each expression is found.

	total mL ×	% saline =	mL of acid
1st solution	8	35%	$0.35 \cdot 8 = 2.8$
2nd solution	x	60%	$0.6x$
Total mixture	$8 + x$	50%	$2.8 + 0.6x$

STEP 2 Write an equation using the information about the total mixture. The chart shows that the total mL $(8 + x)$ multiplied by 50%, or 0.5, is equal to the total mL of acid $(2.8 + 0.6x)$.

$$0.5(8 + x) = 2.8 + 0.6x$$
$$4 + 0.5x = 2.8 + 0.6x$$
$$1.2 = 0.1x$$
$$x = 12$$

STEP 3 Solve for x.

Shayna must add **12 mL** of the 60% acid solution to the beaker.

Solve. A chart is provided for the first two problems.

1. An alloy is made by combining metal A with metal B. Metal A costs $7 per ounce (oz), and metal B costs $10 per oz. The alloy costs $8 per oz. How many ounces of each metal are needed to make 15 oz of the alloy?

	ounces ×	price =	value
Metal A	a	7	7a
Metal B	b	10	10b
Alloy	15	8	120

2. Nita wants to make 10 quarts (qt) of a 72% sugar solution by mixing together a 90% sugar solution and a 70% sugar solution. How much of each solution should she use?

	total qt ×	% sugar =	qt of sugar
Sol 1	x	90%	
Sol 2	y	70%	
Total	10	72%	

3. A landscaper needs 9 cubic feet (ft³) of soil containing 80% clay. If he combines soil with 40% clay and pure clay, how many cubic feet of each soil should he use?

(**Hint:** Pure clay is 100% clay.)

4. In chemistry class, Lisa needs to make 50 mL of a 55% saline solution by mixing together a 65% saline solution and a 40% saline solution. How many milliliters of each solution should she use?

5. How many kilograms (kg) of brand A coffee, at $19 per kg, must be added to 1 kg of brand B coffee, at $14 per kg, to make Carlo's special coffee blend valued at $18 per kg?

6. How much of brand X fruit punch (20% real juice) must be mixed with 2 liters (L) of grape juice (pure juice) to create a mixture containing 52% fruit juice?

7. If 6 fluid ounces (fl oz) of a 55% saline solution is mixed with 9 fl oz of a 25% saline solution, what is the concentration of the new mixture?

(**Hint:** The concentration is the percent of saline in the new mixture.)

8. A gardener mixed 2 cubic feet of soil containing 40% sand with 3 cubic feet of soil containing 50% sand. What is the sand content of the mixture?

9. Twenty pounds (lb) of mixed nuts containing 20% peanuts were mixed with 12 lb of peanuts. Peanuts are what percent of the new mixture?

10. How many kilograms of cane molasses, at $1 per kg, must be added to 2 kg of beet molasses, at $4 per kg, to make Jayton's Premium Molasses, which cost $2 per kg?

11. Pranav wants to make a 12% alcohol solution. He has already poured 9 fl oz of pure water into a beaker. How many fluid ounces of a 30% alcohol solution must he add to the beaker to create the desired mixture?

12. How many kilograms of Indonesian cinnamon, at $17 per kg, must be added to 14 kg of Thai cinnamon, at $20 per kg, to make Perfect Brand Cinnamon, which costs $19 per kg?

A **work problem** is a common algebraic application. In a typical work problem, there is a job that needs to be done. You are told how long it will take a certain number of workers to do the job. Then you are asked to figure out how long the job will take if the number of workers increases or decreases.

Think about your own life experiences. If you have to do some kind of work, the work is finished more quickly if someone helps. In other words, as the number of workers increases, the number of hours the job will take decreases.

To solve a work problem, remember that *work rate × time worked = work done.*

Work rate is how much work a person can do in one hour. Look for a pattern in the following examples:

- Erika can do a job in 2 hours, so she can do $\frac{1}{2}$ the job in one hour.

- Porter can do a job in 6 hours, so he can do $\frac{1}{6}$ the job in one hour.

- Max can do a job in x hours, so he can do $\frac{1}{x}$ the job in one hour.

EXAMPLE 3   Julie can finish a job in 4 hours. Tyler can finish the same job in 6 hours. If they work together, how many hours would it take them to do the job together?

STEP 1   Find the work rate for both workers:

Julie can do $\frac{1}{4}$ the job in one hour, and Tyler can do $\frac{1}{6}$ the job in one hour.

STEP 2   Write an equation. Julie and Tyler's combined work rate multiplied by the time they work must equal 1 job: $\left(\frac{1}{4} + \frac{1}{6}\right)t = 1$.

STEP 3   Solve the equation.

Clear the fractions from the equation by multiplying both sides by the lowest common denominator (LCD) of all the fractions.

$$\left(\frac{1}{4} + \frac{1}{6}\right)t = 1$$
$$\frac{t}{4} + \frac{t}{6} = 1$$
$$12\left(\frac{t}{4} + \frac{t}{6}\right) = 12(1)$$
$$3t + 2t = 12$$

STEP 4   The answer represents time, so express it as hours: $\frac{12}{5} = 2.4$ **hours.**

$$5t = 12$$
$$t = \frac{12}{5}$$

One of the best ways to check your work is to use your common sense. Does the answer make sense? Julie could have done the job in 4 hours. With help, she should be able to get the job done more quickly. An answer close to $2\frac{1}{2}$ hours makes sense.

You can also solve for the work rate of one of the workers. Study the following example and see how it relates to the formula: *work rate × time worked = work done.*

EXAMPLE 4 Ryan can paint a room in 10 hours. If Ashley helps, they can paint the room in 6 hours. How many hours would it take Ashley to paint the room alone?

STEP 1 Find the work rate for both workers. The work rate equals the number of hours the person worked divided by the number of hours it would take that worker to do the job alone. The work rate represents the fraction of the job that the worker completed.

- Ryan's work rate is $\dfrac{\text{hours worked}}{\text{hours to do the job alone}} = \dfrac{6}{10} = \dfrac{3}{5}$.

- Ashley's work rate is $\dfrac{\text{hours worked}}{\text{hours to do the job alone}} = \dfrac{6}{x}$.

STEP 2 Write an equation. Set the combined rates equal to 1, because together the workers completed the entire job. $\dfrac{3}{5} + \dfrac{6}{x} = 1$

STEP 3 Solve the equation.

Multiply both sides by the LCD, $5x$.

Ashley could paint the room alone in **15 hours.**

$$\frac{3}{5} + \frac{6}{x} = 1$$
$$5x\left(\frac{3}{5} + \frac{6}{x}\right) = 5x \cdot 1$$
$$3x + 30 = 5x$$
$$30 = 2x$$
$$x = \mathbf{15}$$

Solve.

13. Working alone, Jim can unload a truck in 10 hours. Zach can do the same job in 14 hours. If they work together, how many hours will it take Jim and Zach to unload the truck?

14. Hose A can fill a barrel with oil in 2 hours. Hose B can do the job in 30 minutes. How many minutes will it take to fill the barrel if both hoses are used at the same time?

    (**Hint:** Convert 2 hours to minutes.)

15. A pool has two drains. The first drain can empty the pool in 3 hours. The second drain can empty the pool in 6 hours. If both drains are open at the same time, how many hours will it take to empty the pool?

16. Jake can mow a lawn in 90 minutes. Lauren can mow the same lawn in 60 minutes. If they work together, how many minutes will it take to mow the lawn?

17. Pablo estimates that he can complete a plumbing job alone in 20 hours. If his assistant helps, the job will take only 12 hours. How many hours would it take the assistant to do the job alone?

18. Robin and Bavesh can complete a data entry job in 6 hours 40 minutes. If Bavesh works alone, the job will take 10 hours. How long would it take Robin to do the job alone?

    (**Hint:** Express 40 minutes as a fraction of an hour.)

In some work problems, workers work together for a while and then one worker finishes the job alone. To solve this type of problem, use the basic fact: *work rate × time worked = work done.*

EXAMPLE 5   Clay estimates that he can complete a roofing job in 10 hours. Javiar could do the same job in 12 hours. If Javiar helps Clay for 3 hours, how long will it take Clay to finish the job on his own?

A chart can help you organize the facts.

	work rate	× time worked	= work done
Clay	$\frac{1}{10}$	$3 + x$ hours	$\frac{1}{10}(3 + x)$
Javiar	$\frac{1}{12}$	3 hours	$\frac{3}{12} = \frac{1}{4}$

The variable x represents the time Clay worked alone.

The work done by Clay and Javiar must equal one job. Write an equation and solve for x.

The equation is the sum of the work done by each person, set equal to 1.

Multiply by the LCD, 20, to clear the fractions.

After Javiar leaves, Clay will need to work an additional $4\frac{1}{2}$ hours on his own. You could also figure out that Clay's total time worked was $3 + 4\frac{1}{2} = 7\frac{1}{2}$ hours.

$$\frac{1}{10}(3 + x) + \frac{1}{4} = 1$$
$$\frac{3}{10} + \frac{1}{10}x + \frac{1}{4} = 1$$
$$6 + 2x + 5 = 20$$
$$11 + 2x = 20$$
$$2x = 9$$
$$x = 4\frac{1}{2} \text{ hours}$$

Solve.

19. It takes pipe A 15 hours to fill a tank. It takes pipe B 10 hours to fill the same tank. Both pipes run together for 4 hours, and then pipe B is shut off. How long will it take pipe A to finish filling the tank?

20. Susan can do a job in 10 hours. Her assistant Alicia could do the job in 20 hours. If Alicia helps for 2 hours, how long will it take Susan to finish the job alone?

21. A plumbing repair takes Steve and Barbara $2\frac{2}{3}$ hours. Steve could have made the repair alone in 8 hours. How long would it have taken Barbara to make the repair alone?

22. Dion can shingle a roof in 12 hours. It would take Frank 18 hours to do the same job. If they work together, how long will it take them to shingle the roof?

23. Robot A takes 36 minutes to paint an assembly. Robot B can do the same job in 24 minutes. The robots work together for 12 minutes, and then robot B is shut down. How long will it take robot A to finish the job?

24. **Challenge** Lin can do a job in 10 hours that takes Dave 5 hours to complete. Lin and Dave worked together for 3 hours, and then Lin finished the job alone. How many hours in all did Lin work?

Solving Rate-Time-Distance Problems

In this lesson, you will learn to

- Solve rate applications involving motion.
- Solve problems involving wind and water currents.

The problems in this section involve a concept called **uniform motion.** Uniform motion means that the speed of an object or vehicle stays the same over a period of time. Of course, you know from your own experience that uniform motion rarely happens.

When you drive, run, or ride a bike, you are frequently changing speeds. Still, you can probably relate to the idea of averaging a certain speed. Uniform motion problems assume that an average speed is maintained for a period of time.

The formula used to solve motion problems is *distance = rate × time,* commonly written *d = rt.*

Measurement units are important in motion problems. The rate units must match the distance and time units. For example, if the distance is in *miles* and the time is in *hours,* the rate must be in *miles per hour.*

Basic motion problems can be solved using only one variable. More complex questions can require two variables. Remember, you can use two variables to solve a problem as long as you have enough information to write two equations. If you can write only one equation, you must use only one variable.

EXAMPLE 1 John and his family drove to a campground in the mountains in $1\frac{1}{2}$ hours. Because of bad weather, the return trip took 2 hours. If John drove 10 miles per hour slower on the way home, what was John's average speed on both parts of the trip?

STEP 1 Make a drawing of the situation. Drawing a diagram helps you see how the facts of the problem are related. The drawing makes it clear that the distance traveled is equal for both parts of the trip.

$1\frac{1}{2}$ hr r mph

$r - 10$ mph   2 hr

STEP 2 Use the drawing to organize the facts of the problem in a chart.

	rate (mph) ×	time (hr)	= distance (mi)
Going	r	$1\frac{1}{2}$	$\frac{3}{2}r$
Coming back	$r - 10$	2	$2(r - 10)$

Write the distance as the product of rate and time. Write the mixed number $1\frac{1}{2}$ as an improper fraction, $\frac{3}{2}$.

STEP 3 Write an equation. Because the distance is the same going and coming back, the expressions for distance must be equal.

$\frac{3}{2}r = 2(r - 10)$

STEP 4 Solve the equation.

STEP 5 Look back at the chart. The variable r represents the rate on the first part of the trip.

Substitute for r to find the rate on the return trip:
$$r - 10 = 40 - 10 = \textbf{30 mph}$$

$$\tfrac{3}{2}r = 2(r - 10)$$
$$\tfrac{3}{2}r = 2r - 20$$
$$3r = 4r - 40$$
$$-r = -40$$
$$r = \textbf{40 mph}$$

Make sure you answer the question posed by the problem. **John's average speed was 40 mph on the first part of the trip and 30 mph on the second part.**

 Think About It: Now that you know the rate, could you find the actual number of miles that John drove? You could. Substitute 40 for r in either distance expression. The trip was 60 miles in each direction.

$$\tfrac{3}{2}r = \tfrac{3}{2}(40)$$
$$= \textbf{60 miles}$$

In the next example, time (t) is the unknown variable.

EXAMPLE 2 A cargo ship left port at 7 A.M. and traveled west at an average speed of 13 mph. A cruise ship left 5 hours later and traveled in the same direction but with an average speed of 23 mph. At what time did the cruise ship catch up to the cargo ship?

STEP 1 Make a drawing.

Left at 7 A.M. $r = 13$ mph →

Left 5 hr later $r = 23$ mph →

You can see that distance is equal in this problem. The rates are given. The amount of time is the unknown.

STEP 2 Organize the facts of the problem in a chart.

	rate (mph) ×	time (hr)	= distance (mi)
Cargo ship	13	t	$13t$
Cruise ship	23	$t - 5$	$23(t - 5)$

If the cruise ship left 5 hours later, why is its time written as $t - 5$ instead of $t + 5$?

You subtract 5 because the cruise ship will have traveled 5 fewer hours than the cargo ship.

STEP 3 Write an equation. At the point where the cruise ship catches up to the cargo ship, both ships will have traveled the same number of miles.

$$13t = 23(t - 5)$$
$$13t = 23t - 115$$
$$-10t = -115$$
$$t = \textbf{11.5 hr}$$

STEP 4 Solve for t.

STEP 5 Refer back to the chart. The cargo ship had been traveling for 11.5 or $11\tfrac{1}{2}$ hours. Starting at 7 A.M., count forward $11\tfrac{1}{2}$ hours. **The cruise ship caught up to the cargo ship at 6:30 P.M.**

Solve. A chart is partially completed for selected problems.

1. At 9 A.M., a train leaves Orson traveling north at 90 kilometers per hour (km/h). An hour later, another train leaves Orson traveling north at 120 km/h on a parallel track. How many kilometers from Orson will the trains be when they meet?

	rate ×	time =	distance
Train 1	90	t	
Train 2	120	$t - 1$	

2. Grace had been driving at a constant rate of speed for 3 hours until traffic caused her to reduce her speed by 10 miles per hour for the remaining 2 hours of her 280 mile trip. What was her speed for the first three hours?

	rate ×	time =	distance
1st part	r		
2nd part	$r - 10$		

3. Ken and Sean are on bicycles. They start 60 kilometers apart and begin riding toward each other at 2 P.M. They meet $1\frac{1}{2}$ hours later. Ken's average speed is 4 km/h faster than Sean's average speed. Find Sean's average speed.

(**Hint:** Together, the cyclists ride a total of 60 kilometers.)

	rate ×	time =	distance
Ken	$r + 4$	1.5	
Sean	r		

4. Riding a bicycle on a steep uphill course, Marco averages 5 mph. Riding back down the course, he averages 30 mph. If the trip up and down the hill took 2 hours 20 minutes, how many miles did Marco travel in all?

(**Hint:** 2 hr 20 min equals $2\frac{1}{3}$ hr.)

	rate ×	time =	distance
Uphill		t	
Downhill		$2\frac{1}{3} - t$	

5. Rosa begins running on a jogger's trail at 8 A.M. She runs at an average rate of 5 mph. Ten minutes later, Sandi begins running the same trail. How fast must Sandi run to catch up to Rosa by 8:30 A.M.?

(**Hint:** Change minutes to fractions of an hour.)

	rate ×	time =	distance
Rosa			
Sandi			

6. A jet took off from an airport at 2 P.M. and flew south. A second jet took off one hour later and flew north at a speed 140 km/h faster than the speed of the first jet. At 6 P.M., the jets were 1,918 kilometers apart. What was the speed of the first jet?

	rate ×	time =	distance
1st jet			
2nd jet			

Students are sometimes surprised to discover that they can add or subtract from a rate. Think about it this way: Suppose you are on a cruise ship that is traveling north at 20 mph. You decide to jog north on the deck of the ship. If your average jogging speed is 4 mph, how fast are you traveling north? The answer is 24 mph. As you jog north, you are actually traveling faster than the ship, which is why you will eventually arrive at the front of the ship.

In the next set of problems, you will explore the effects of water and air currents on travel. Moving with a current increases speed, adding to the rate, while moving against a current slows you down, subtracting from the rate.

EXAMPLE 3 A boat traveled 80 miles downstream and then traveled back. The trip downstream took 4 hours. The trip back took 8 hours. Find the speed of the boat in still water and the speed of the current.

STEP 1 Make a drawing.

- When the boat goes downstream, the current adds to the speed.

Downstream — Boat speed r + current c → 4 hr trip

Upstream ← Boat speed r − current c — 8 hr trip

- When the boat goes upstream, against the current, the current is subtracted from the speed.

STEP 2 Organize the facts of the problem in a chart.

	rate (mph) ×	time (hr)	= distance (mi)
Downstream	$r + c$	4	$4(r + c)$
Upstream	$r - c$	8	$8(r - c)$

The problem states that the distance one way is 80 miles.

There are two expressions in the chart that equal distance.

STEP 3 Write a system of equations.

$$\begin{cases} 4(r + c) = 80 \\ 8(r - c) = 80 \end{cases}$$

STEP 4 Solve for r and c.

Multiply to clear the parentheses.

$$4r + 4c = 80$$
$$8r - 8c = 80$$

> **Think About It:** Because the distance is the same coming and going, you may have considered writing this equation:
>
> $4(r + c) = 8(r - c)$
>
> Why wouldn't that help?

Use the elimination method. Multiply both sides of the first equation by 2.

$$8r + 8c = 160$$
$$\underline{8r - 8c = 80}$$
$$16r = 240$$
$$r = \textbf{15}$$

Substitute 15 for r in the first equation and solve for c.

$$4r + 4c = 80$$
$$4(15) + 4c = 80$$
$$60 + 4c = 80$$
$$4c = 20$$
$$c = \textbf{5}$$

STEP 5 Refer back to the chart. The speed of the boat in still water (r) is **15 mph**. The speed of the current (c) is **5 mph**.

Solve. A chart is partially completed for selected problems.

7. A boat traveled 462 miles downstream and then traveled back. The trip downstream took 14 hours. The trip back took 22 hours. What was the speed of the boat in still water? What was the speed of the current?

	rate ×	time =	distance
Down	r + c	14	
Up	r − c	22	

8. A plane traveled 1,400 miles each way to Toronto and back. Flying with the wind, the plane trip took 7 hours. The trip home, into the wind, took 10 hours. What was the speed of the plane in still air? What was the speed of the wind?

	rate ×	time =	distance
With	r + c	7	
Against	r − c	10	

9. A boat traveled 168 miles downstream and then traveled back. The trip downstream took 6 hours. The trip back took 12 hours. What was the speed of the current?

	rate ×	time =	distance
Down			
Up			

10. A jet flew 2,400 miles with the wind in 12 hours. The return trip against the wind took 15 hours. What was speed of the jet in still air? What was the speed of the wind?

	rate ×	time =	distance
With			
Against			

The next two problems are variations of the type you have been doing. Motion problems will not always have the same wording, but they are always based on the same three quantities—rate, time, and distance. Find a way to represent these quantities, and then set up an equation or a system of equations.

11. Motu is planning a three-hour canoe trip down a river and back. He knows that he can paddle in still water at 4 mph. He also knows that the rate of the current is 2 mph. How much time can he spend going downstream before he needs to turn back? How far downstream will he have traveled?

	rate ×	time =	distance
Down		t	
Up		3 − t	

12. A motorboat can travel 30 km/h in still water. If the boat can travel 432 km downstream in the same amount of time that it can travel 288 km upstream, what is the speed of the current?

(**Hint:** If $d = rt$, then $t = \frac{d}{r}$.)

	rate ×	time =	distance
Down			
Up			

Solving Geometry Applications

In this lesson, you will learn to

- Solve area and perimeter problems.
- Solve similar figure problems.

Algebra problems are often based on geometric figures. One common type of problem gives the area and the perimeter of a figure and asks you to find the length and the width (or height).

You know that the perimeter of a figure is the sum of the lengths of the sides. Formulas exist for finding the perimeter of various shapes, but the concept remains the same. To find the perimeter, add all the sides.

Area is a measure of the space inside a 2-dimensional figure. Remember that area must always be measured in square units. Area formulas are helpful. You should master the following formulas:

AREA FORMULAS	
Area of a Rectangle: *length × width*	$A = lw$
Area of a Square: *side × side*	$A = s^2$
Area of a Parallelogram: *base × height*	$A = bh$
Area of a Triangle: $\frac{1}{2} \times$ *base × height*	$A = \frac{1}{2}bh$

Can you see how these formulas are related? In some way, they all multiply *base* and *height*.

EXAMPLE 1 A rectangle is 5 times as long as it is wide. If its length was decreased by 24 inches and its width was increased by 24 inches, the figure would be a square. What are the dimensions of the rectangle?

STEP 1 Let l represent length and w represent width.

Make a sketch of the figures in the problem.

$l - 24$

$w + 24$

l

w

If you aren't given a drawing, always make one, even if the problem seems simple. A sketch helps you focus on the right details.

STEP 2 Write a system of equations. You know the length of the rectangle is 5 times the width. For the second equation, use the fact that the sides of a square are equal.

STEP 3 Solve the system. Substitute $5w$ for l in the second equation.

STEP 4 Go back to the first equation to find l.

The dimensions of the rectangle are **60 inches by 12 inches.**

$$\begin{cases} l = 5w \\ l - 24 = w + 24 \end{cases}$$

$$5w - 24 = w + 24$$
$$4w = 48$$
$$w = \mathbf{12}$$
$$l = 5w$$
$$l = 5(12)$$
$$l = \mathbf{60}$$

When you are working with area, one of your equations may contain a variable raised to the second power. Study the next example to see how to find the value of the variable.

EXAMPLE 2   The area of a parallelogram is 324 square inches. The base of the parallelogram is 4 times its height. Find the length of the base.

STEP 1 Let b represent the base and h represent the height.

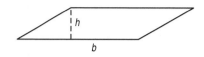

Make a drawing.

STEP 2 Write a system of equations. Use the formula for the area of a parallelogram to write one of the equations.

$$\begin{cases} b = 4h \\ bh = 324 \end{cases}$$

STEP 3 Solve the equation using substitution. Substitute the value $4h$ for b in the second equation.

$$4h \cdot h = 324$$
$$4h^2 = 324$$

Divide both sides by 4.

$$h^2 = 81$$
$$\sqrt{h^2} = \sqrt{81}$$

Take the square root of both sides.

$$h = 9$$

Note: You can ignore the solution −9 since the height of a figure must be a positive number.

STEP 4 Go back to the first equation and solve for b.

The length of the base is **36 inches**.

$$b = 4h$$
$$b = 4(9)$$
$$b = 36$$

Solve the following problems by writing a system of equations.

1. The length of a rectangle is 6 cm greater than its height. If you decrease the length by 10 cm and increase the width by 10 cm, the perimeter of the new rectangle is 132 cm. Find the length and width of the original rectangle.

2. A rectangle has an area of 800 square centimeters. If the length of the rectangle is twice its width, what is the perimeter of the rectangle?

3. The area of a right triangle is 36 square inches. If the height of the triangle is twice its base, what is the length of the base?

4. A rectangle has a perimeter of 80 feet. If you subtract 5 feet from its length and add 5 feet to its width, the new figure is a square. What is the area of the new square?

5. A rectangular garden has an area of 128 square yards. If its width is half as long as its length, what is the perimeter of the garden?

6. The length of a rectangle is 1 inch greater than twice its width. If you decrease both the length and the width by 2 inches, the new perimeter will be 24 inches. Find the length and width of the original rectangle.

7. The perimeter of an isosceles triangle is 42 inches. Its base is 6 inches less than the sum of its other legs. Find the lengths of the three sides.

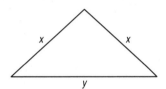

8. **Challenge** On a coordinate plane, a triangle is formed by the x-axis and the graphs of the equations $-x + y = 1$ and $2x + y = 10$. What is the area of the triangle in square units?

(**Hint:** The length of the base is the distance between the x-intercepts. The height is the y-coordinate of the point where the lines intersect.)

Synthesis Some geometry applications use proportional reasoning. In geometry, two figures are said to be **similar** if all corresponding angles are equal. If the corresponding angles are equal, then the lengths of the corresponding sides will be proportional.

You know the triangles to the right are similar (~) because the angles are equal.

$\triangle ABC \sim \triangle DEF$

You can also tell they are similar because the lengths of the corresponding sides are proportional.

$\dfrac{AB}{DE} = \dfrac{BC}{EF} = \dfrac{CA}{FD}$

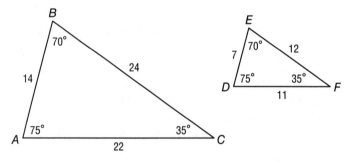

The scale factor of $\triangle ABC$ to $\triangle DEF$ is 2 to 1.

Two triangles are similar if *any* of the following conditions are met:

AA	Angle-Angle	Two angles in one triangle are equal to two angles in another.
SSS	Side-Side-Side	The lengths of the sides in one triangle are proportional to the sides in another triangle.
SAS	Side-Angle-Side	One angle in one triangle is equal to one angle in the other triangle, and the sides that form those angles are proportional.

You can solve for missing sides or angles algebraically.

EXAMPLE 3   If side *AC* measures 36 cm, what is the value of *x* and *y* in the figure?

STEP 1   Find the similar triangles. Triangles *ABE* and *ACD* are similar because they have two equal angles. ∠*ABE* and ∠*ACD* both measure 86°, and ∠*CAD* and ∠*BAE* both measure 35°.

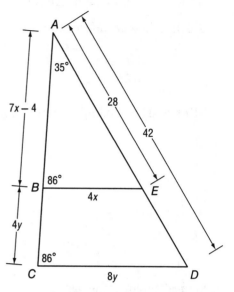

STEP 2   You know that corresponding sides are proportional. Write a statement of proportionality.

$$\frac{AE}{AD} = \frac{BE}{CD} = \frac{AB}{AC}$$

STEP 3   Use the proportions and the facts from the problem to write equations.

Equation 1: $\frac{AE}{AD} = \frac{BE}{CD}$ so $\frac{28}{42} = \frac{4x}{8y}$

Simplify and cross multiply. Then rewrite in standard form:

$$\frac{\overset{2}{\cancel{28}}}{\underset{3}{\cancel{42}}} = \frac{\overset{1}{\cancel{4x}}}{\underset{2}{\cancel{8y}}} \quad \text{becomes} \qquad 4y = 3x$$
$$-3x + 4y = 0$$

> **Tip:** When a drawing has overlapping triangles, you may find it helpful to redraw the triangles so that they appear side by side. Take time to make sure that every side and angle is labeled correctly.

Equation 2: $AB + BC = 36$
$(7x - 4) + 4y = 36$
$7x + 4y = 40$

STEP 4   Put the equations together to write and solve a system:

$$\begin{cases} -3x + 4y = 0 \\ 7x + 4y = 40 \end{cases}$$

Multiply the terms in the first equation by −1 and add to eliminate the variable *y*.

$$\begin{aligned} 3x - 4y &= 0 \\ 7x + 4y &= 40 \\ \hline 10x &= 40 \\ x &= 4 \end{aligned}$$

Go back to the original equation and substitute 4 for *x*.

$$\begin{aligned} -3(4) + 4y &= 0 \\ -12 + 4y &= 0 \\ 4y &= 12 \\ y &= 3 \end{aligned}$$

The solution is **x = 4** and **y = 3**.

Now that you know the values of *x* and *y*, you could answer many other questions about the diagram. Always make sure that you answer the question asked in the original problem.

Try these questions:

Answers:

- What is the measure of side *BE*?

  16 cm
- Find the measure of ∠*AEB*.

  59°
- How much greater is the perimeter of △*ACD* than that of △*ABE*?

  34 cm

Solve by writing a system of equations.

9. Triangles *JKN* and *LMN* are similar triangles.
   What is the perimeter of △*JKN*?

   Use the following information:

   $$KM = 16$$
   $$MN = 2y + 4$$

   (**Hint:** Start by finding the ratio $\frac{LN}{JN}$.)

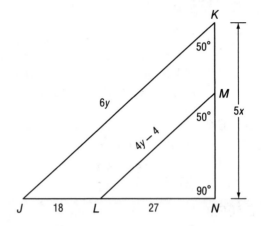

10. In the diagram at the right,
    all units are in centimeters.

    a. Find the value of *x* and *y*.

    b. What is the length of
    side *AE*?

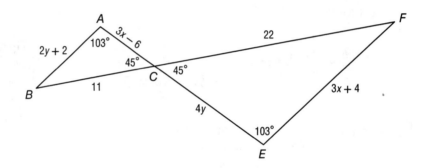

11. Triangles *QPS* and *RQS* are similar. Side *PS* of
    △*QPS* corresponds to side *QS* of △*RQS*.

    The ratio of side *PQ* to side *QS* is 3 to 4, and the
    length of side *PS* is 20 inches.

    a. Find the perimeter of △*QPS*.

    b. **Challenge** Find the perimeter of △*RQS*.

$$PS = \frac{x+y}{2} + 6$$

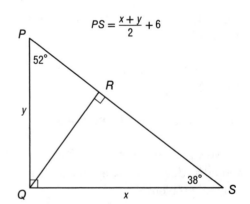

Problem Solving with Multiple Variables Review

Solve the problems below. When you finish, check your answers at the back of the book and correct any errors.

Solve each system by graphing.

1. $6x + 7y = 7$
$x + 7y = -28$

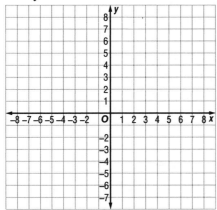

$x - y = -5$
$y = 6$

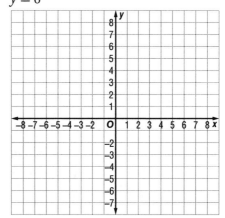

2. $-5x + 4y = 12$
$x + y = -6$

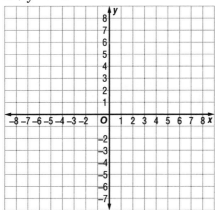

$x - 4y = -12$
$x + y = -2$

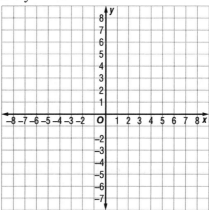

3. $3x + y = 4$
$2x - 3y = 21$

$7x - 4y = 32$
$x = 8$

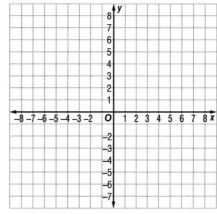

Solve each system using the method of your choice. Write *inconsistent* or *dependent* if the system does not have a single solution.

4. $y = -4x + 1$
   $6x + 2y = -2$

   $8x + 7y = -13$
   $y = -3x$

   $7x - 5y = 19$
   $x - 3y = 21$

5. $-3x - 4y = 19$
   $x - 7y = 2$

   $8x - 3y = 8$
   $2x + y = 16$

   $y = 7x + 1$
   $-21x + 3y = -6$

6. $-7x - 3y = 9$
   $-2x + y = 10$

   $5x + y = 16$
   $-10x - 2y = -32$

   $-5x - y = -5$
   $x - 4y = -20$

7. $16x - 3y = -10$
   $8x - 4y = 0$

   $2x + y = 20$
   $x + 2y = 13$

   $-5x - y = 22$
   $-7x - 5y = 2$

8. $-8x + 5y = 26$
   $3x + 4y = 2$

   $-3x - 3y = -9$
   $4x + 4y = 12$

   $-5x - 8y = -23$
   $2x - 3y = 3$

9. $4x - 20y = -16$
   $-18x + 90y = 0$

   $-3x + 2y = -27$
   $-4x - 3y = -2$

   $-5x - 6y = 2$
   $9x + 7y = -15$

A student made an error when solving the following system: $2x - y = -2$
$x - 4y = -15$

Circle the step that contains the student's mistake. Explain the mistake.

10. A. Rewrite one equation to solve for x.

   B. Go back to the original system and substitute $4y - 15$ for x.

   C. Solve for y.

   $2x - y = -2$
   $x = 4y - 15$

   $(4y - 15) - 4y = -15$

   $4y - 15 - 4y = -15$
   $-15 = -15$

   Student's answer:
   The system is dependent.

Solve.

11. A fair charges $4.50 for children and $9 for adults. On the weekend, 2,400 people attended the fair and $14,715 was collected in ticket sales. How many children's tickets were sold?

12. The sum of the digits of a certain two-digit number is 12. When you reverse its digits, you increase the number by 54. What is the number?

13. Yam and Silas are selling small and large boxes of cookies for a school fund-raiser. Yam sold 20 small and 10 large boxes for a total of $290. Silas sold 40 small and 35 large boxes for a total of $775. Find the cost of a small box of cookies.

14. Donna is 4 years older than her twin brothers Justin and Jared. Next year the sum of all of their ages will be 40. How old is Jared now?

15. If 1 is subtracted from the denominator of a fraction, the value of the fraction is $\frac{2}{3}$. If 4 is added to the numerator of the original fraction, the value of the fraction is $\frac{3}{4}$. What is the original fraction?

16. A fraction can be simplified to $\frac{2}{3}$. In the original fraction, the numerator is 6 less than the denominator. What is the denominator of the original fraction?

Use the facts to find the constant k. Then write the equation of variation. If necessary, express k as a decimal.

17. y varies directly as x, and $y = 52$ when $x = 13$

y varies directly as x, and $y = 40$ when $x = 25$

18. y varies inversely as x, and $y = 9$ when $x = 3$

y varies inversely as x, and $y = 14$ when $x = 2$

19. y varies directly as x, and $y = 16$ when $x = 5$

y varies inversely as x, and $y = 18$ when $x = 5$

Solve.

20. A bag has green and white marbles. There are 3 green marbles for every 7 white ones. If you add 2 green marbles and take out 2 white ones, there will be 1 green marble for every 2 white ones. How many green and white marbles are in the bag originally?

21. One number is 16 more than another. If the smaller number is subtracted from $\frac{3}{4}$ of the larger number, the result is $\frac{1}{8}$ of the sum of the two numbers. What are the numbers?

22. Daniel has only dimes and nickels in his pocket. The ratio of dimes to nickels is 7 to 4. If Daniel removes 5 dimes and adds 5 nickels, the ratio of dimes to nickels will be 6 to 5. How many dimes and nickels are in Daniel's pocket originally?

23. At a restaurant, the ratio of customers who ordered the chef's special to those who ordered from the menu was 2 to 1. If 10 customers change their order from the menu to the chef's special, the ratio will be 3 to 1. How many customers are at the restaurant?

24. A landscaper needs 11 cubic meters (m³) of soil containing 52% clay. She combines soil A with 34% clay and with soil B, which is pure clay. How many cubic meters of each soil should she use?

(**Hint:** Pure clay is 100% clay.)

25. Lopez Market sells 32-ounce bags of mixed nuts that contain 35% peanuts. To make the product, Lopez Market combines Sun Valley Nuts, which is 50% peanuts, and Rancho mixed nuts, which is 30% peanuts. How much of each nut mix is used?

26. Scott needs to make a 71% sugar solution. He has already placed 12 drops of a 75% sugar solution into a test tube. How many drops of a 55% sugar solution should he add to the test tube to create the desired solution?

27. Alma mixes 8 fl oz of an alcohol solution with 2 fl oz of pure water to make a 24% alcohol solution. What was the alcohol concentration in the first solution?

(**Hint:** Pure water is 0% alcohol.)

28. A metal alloy weighing 4 kg and containing 48% tin is melted and mixed with 12 kg of a different alloy that contains 84% tin. What percent of the resulting alloy is tin?

29. Marshea needs to make 16 ounces of an alloy containing 97% iron. She is going to melt and combine one metal that is 88% iron with pure iron. How many ounces of each should she use?

30. Together, Logan and Jayden can mow a lawn in 10 minutes. Logan can mow the lawn alone in 15 minutes. How long would it take Jayden to mow the lawn by himself?

31. Pipe A can fill a large tank in 24 hours. Together, pipe A and pipe B can fill the same tank in 8 hours. How long would it take to fill the tank using only pipe B?

32. Two painters, working together, can finish a job in 9 hours. Working separately, one of the painters could do the job in 12 hours. How long would it take the other painter to do the job alone?

33. Melanie can complete 40 patient files in 15 hours. If Eric helps her, the job will take only 6 hours. How long would it take Eric to complete the files alone?

34. T&R Construction can complete a road construction project in 6 days. Case Asphalt can complete the same project in 10 days. How long would it take the two companies, working together, to complete the project?

35. A large pool has two drains. The pool can be drained completely by drain A in 15 hours or by drain B in 45 hours. How long will it take to drain the pool if both drains are opened?

36. Trains A and B left the station at the same time and traveled in opposite directions. Train A traveled 15 kilometers per hour faster than train B. After 18 hours, the trains were 2,070 km apart. What was the average speed of train B?

37. A diesel train left Eastwood and traveled north at an average speed of 50 mph. A passenger train left one hour later and traveled in the same direction but with an average speed of 60 mph. How long did the diesel train travel before the passenger train caught up?

38. A boat traveled 576 miles downstream and then traveled back. The trip downstream took 16 hours. The trip back took 32 hours. Find the speed of the boat in still water and the speed of the current.

39. A cruise ship traveled 120 miles south with the current for 6 hours. The return trip north took 12 hours. What was the speed of the current?

40. In a rectangle, the ratio of the length to the width is 4 to 3. If you subtract 6 cm from the length and add 6 cm to the width, the resulting figure is a square. Find the perimeter of the rectangle.

41. The length of a rectangle is 4 inches greater than twice its width. If you divide both the length and the width by 2, the perimeter of the rectangle is 58 inches. Find the length and width of the original rectangle.

42. An isosceles triangle has two legs of the same length and a base. The ratio of the base to one leg is 3 to 2. If the perimeter of the triangle is 35 cm, what are the lengths of the base and the legs?

43. A rectangular lot has an area of 588 square yards. If the width of the lot is $\frac{1}{3}$ its length, what is the perimeter of the lot?

44. The area of a right triangle is 96 square inches. If the base of the triangle is three times its height, what is the height of the triangle?

45. The length of a rectangle is 3 meters more than twice its width. If the perimeter of the rectangle is 42 meters, what are the dimensions of the rectangle?

46. Rectangles *ABCD* and *WXYZ* are similar.

The ratio of side *AB* to side *WX* is 5 to 4.

The ratio of side *AB* to side *AD* is 5 to 3.

If side *XY* measures 7.2 inches, what is the area of rectangle *ABCD*?

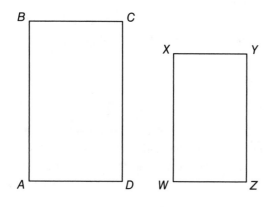

47. Find the area, in square units, enclosed between the *y*-axis and the graphs of the equations $y = x - 4$ and $y = -x + 2$.

(**Hint:** Think of the *y*-axis as the base of the triangle.)

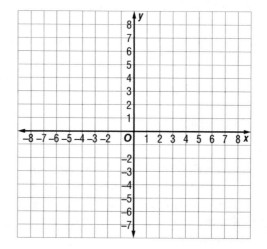

QUADRATIC EQUATIONS

You have been working with equations that contain two variables, x and y. In this section you will explore what happens when the variable x is raised to the second power.

Basics of Quadratic Equations

In this lesson, you will learn to

- Recognize quadratic equations.

- Write quadratic expressions in standard form.

The graph of a linear equation is a straight line. Why does that happen? In a linear equation, both variables x and y are raised to the first power. (Remember, $x = x^1$ and $y = y^1$.) Every point on the line has x- and y-coordinates that are a solution to the equation. This progression of solutions creates a pattern of a straight line.

In a **quadratic equation,** one of the terms includes x raised to the second power.

Consider the equation $x^2 = 4$.

What value of x makes the equation true?

You might immediately say that x equals 2 because $2^2 = 4$, but there is another possible answer. The variable x could equal -2 because $(-2)^2 = 4$. Can you see why the variable x could have more than one value in a quadratic equation? This is because two different x-values can yield the same result.

Think back to the linear equations you solved in the last section. A line forms from the solutions because there are two variables, x and y. Without two variables, the solutions couldn't be plotted on a coordinate plane with x- and y-axes.

With that in mind, what happens when you graph a quadratic equation with two variables? What shape does the solution set form?

Look at the tables of values for two equations, one linear and one quadratic.

Linear: $y = x$

Each y has one and only one possible x.

x	y
0	0
1	1
2	2
3	3

Quadratic: $y = x^2$

Here the same y-value can be produced by two different x-values.

x_1	x_2	y
0	0	0
1	−1	1
2	−2	4
3	−3	9

The table for the linear equation shows the following coordinates: $(0, 0)$, $(1, 1)$, $(2, 2)$, and $(3, 3)$.

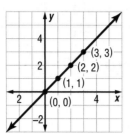

Linear equation $y = x$

The table for the quadratic equation shows these ordered pairs: $(0, 0)$, $(1, 1)$, $(-1, 1)$, $(2, 4)$, $(-2, 4)$, $(3, 9)$, and $(-3, 9)$.

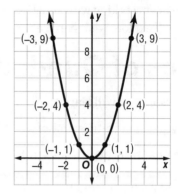

Quadratic equation $y = x^2$

The graph of the quadratic equation is a **parabola.** You will learn more about this unique shape when you study graphing quadratics. For now, make sure you understand why the solutions to the quadratic equation create this shape instead of a straight line.

The **roots** of a quadratic equation are the solutions, or values of x, when y is 0.

What do these roots look like on a graph?

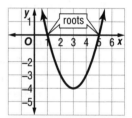

The parabola at the right is the graph of the equation $y = x^2 - 6x + 5$. The roots are the points where the parabola crosses the x-axis.

You can see that the roots $(1, 0)$ and $(5, 0)$ are solutions to the equation $x^2 - 6x + 5 = 0$.

One of the strategies for finding the roots of a quadratic equation is to replace y with 0 and solve for x.

Before solving any quadratic equation, first write the equation in standard form.

The **standard form** of a quadratic equation is written $ax^2 + bx + c = 0$, where a, b, and c are real numbers, and $a \neq 0$.

 Think About It: Do you see why a cannot equal 0? If a equals 0, then ax^2 equals 0, leaving the equation $bx + c = 0$, which is not a quadratic equation.

You must be able to write a quadratic equation in standard form and determine the values of a, b, and c. Notice that the terms are written so that the powers of the variable x are in descending order.

EXAMPLE 1   Write $10x = 9 - 12x^2$ in standard form, and find a, b, and c.

STEP 1   Use inverse operations to get all terms on the left side, and set the left side equal to zero.

$$10x = 9 - 12x^2$$
$$12x^2 + 10x - 9 = 0$$

STEP 2   Find the values of a, b, and c.

$$a = 12$$
$$b = 10$$
$$c = -9$$

EXAMPLE 2   Write $3x^2 = -2x$ in standard form, and find a, b, and c.

STEP 1   Use inverse operations to get all terms on the left side, and set the left side equal to zero.

$$3x^2 = -2x$$
$$3x^2 + 2x = 0$$

STEP 2   Find the values of a, b, and c.

Since there is not a third term, c is equal to 0.

$$a = 3$$
$$b = 2$$
$$c = 0$$

Write each quadratic equation in standard form, and find the values of a, b, and c.

1. $7x^2 = 5$ $x^2 - 4x = 12$ $9x^2 = 20 + 6x^2$

$a =$ $a =$ $a =$
$b =$ $b =$ $b =$
$c =$ $c =$ $c =$

2. $8x = 48 - x^2$ $4x^2 = 8$ $x^2 + 21 = 10 + 12x$

$a =$ $a =$ $a =$
$b =$ $b =$ $b =$
$c =$ $c =$ $c =$

3. $x^2 - 4x - 21 = -3 + 4x$ $x - 120 = -2x^2$ $x^2 = 11 - 7x$

$a =$ $a =$ $a =$
$b =$ $b =$ $b =$
$c =$ $c =$ $c =$

4. $3x^2 - 5x = 5 - 2x$ $x - 34 = -11x^2 - 10$ $9x^2 = 8 + 6x^2$

$a =$ $a =$ $a =$
$b =$ $b =$ $b =$
$c =$ $c =$ $c =$

Solve Quadratic Equations by Factoring

In this lesson, you will learn to

- Find the roots of a quadratic equation when the factorization is given.

- Solve quadratic equations by factoring.

You can find the roots of a quadratic equation by setting the equation equal to zero. You will need to apply the **zero-product principle.**

Zero-product principle	If $a \cdot b = 0$, then $a = 0$ or $b = 0$, or both.

In other words, if you know that the product of two factors is 0, then one or both of the factors must be 0.

EXAMPLE 1 The quadratic equation $x^2 + 5x - 6 = 0$ can be written as $(x - 1)(x + 6) = 0$. What are the two solutions?

In order for the equation to equal 0, either $x - 1$ or $x + 6$ must equal 0. Set each factor equal to 0 and solve for x.

$$x - 1 = 0 \qquad x + 6 = 0$$
$$x = 1 \qquad\quad x = -6$$

The two solutions are **1** and **−6.** The solutions can be written in brackets to show they both belong to the set of solutions to this problem: {1, −6}.

Recall from the last section that this means the parabola representing the equation $y = x^2 + 5x - 6$ crosses the x-axis in two places, (1, 0) and (−6, 0).

The solutions to a quadratic equation may be fractions.

EXAMPLE 2 The quadratic equation $12x^2 + 28x + 15 = 0$ can be written as $(2x + 3)(6x + 5) = 0$. What are the two solutions?

Set each factor equal to 0 and solve for x.

The solutions are $\left\{-\frac{3}{2}, -\frac{5}{6}\right\}$.

$$2x + 3 = 0 \qquad 6x + 5 = 0$$
$$2x = -3 \qquad\quad 6x = -5$$
$$x = -\frac{3}{2} \qquad\quad x = -\frac{5}{6}$$

If both factors are the same, there will be only one solution.

EXAMPLE 3 The quadratic equation $x^2 - 8x + 16 = 0$ can be written as $(x - 4)^2 = 0$. Find the solution.

Although it is used twice, there is only one factor. $(x - 4)^2 = 0$ is the same as $(x - 4)(x - 4) = 0$.

$$x - 4 = 0$$
$$x = 4$$

The only solution is **{4}.**

Could x ever equal 0? Your experience with graphing confirms that a line could cross the x-axis at 0. What does a quadratic equation with a root of 0 look like?

EXAMPLE 4 The quadratic equation $x^2 - 9x = 0$ can be written as $x(x - 9) = 0$. Find the solutions.

There are two factors: x and $x - 9$. Set each factor equal to 0, and solve for x.

$$x = 0 \quad x - 9 = 0$$
$$x = 9$$

The solution set is {0, 9}.

Solve the following quadratic equations. The equations are in factored form.

1. $(x + 7)(x + 1) = 0$ $(a - 6)(a + 3) = 0$ $(b + 4)(b - 4) = 0$

2. $(n + 2)(n - 4) = 0$ $(c + 5)^2 = 0$ $(m + 1)(3m - 4) = 0$

3. $x(x - 6) = 0$ $(7n - 1)(n - 5) = 0$ $(6a - 5)(a - 2) = 0$

4. $n(4n + 1) = 0$ $(6x + 1)(8x + 3) = 0$ $(2b - 5)(5b - 4) = 0$

5. $(x - 10)^2 = 0$ $(m - 3)(m + 1) = 0$ $(8n + 3)(2n - 3) = 0$

The next step in solving quadratics is to learn how to write a quadratic equation in factored form. You have already used FOIL to multiply two binomials (page 20). Review the process in the diagram at the right. The diagram shows the multiplication of $(x + 2)(x - 4)$ to get $x^2 - 4x + 2x - 8$, which simplifies to the trinomial $x^2 - 2x - 8$.

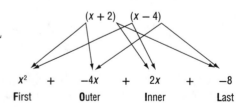

Make sure you understand how each term in the factors is used twice.

To write a quadratic in factored form, you will need to reverse the process. If you are given a quadratic equation in the form of a trinomial, you first need to find the factors. Then you can set the factors equal to 0 and find the solutions to the equation.

The next example shows how to solve a quadratic equation in the form $ax^2 + bx + c = 0$ when $a = 1$. The binomial factors will take the form $(x + \boxed{?})(x + \boxed{?})$, where $\boxed{?} + \boxed{?} = b$ and $\boxed{?} \times \boxed{?} = c$.

As you look at the following examples, pay attention to whether the b and c terms are positive or negative. Then think about what numbers will produce those results.

EXAMPLE 5 Solve: $x^2 + 14x + 48 = 0$

STEP 1 Factor the left side of the equation. You need to find two numbers that multiply to 48 and add to 14. You know that $6 \times 8 = 48$ and $6 + 8 = 14$. Use these numbers to write the binomial factors.

$$x^2 + 14x + 48 = 0$$
$$(x + 6)(x + 8) = 0$$

STEP 2 Set each factor equal to 0 and solve for x.

$$x + 6 = 0 \qquad x + 8 = 0$$
$$x = -6 \qquad x = -8$$

The solution set is $\{-8, -6\}$.

STEP 3 Check your work by substituting each solution into the original equation. Then solve to see if the equation is true.

$$x^2 + 14x + 48 = 0$$
$$(-8)^2 + 14(-8) + 48 = 0$$
$$64 - 112 + 48 = 0$$
$$0 = 0 \text{ True}$$

$$x^2 + 14x + 48 = 0$$
$$(-6)^2 + 14(-6) + 48 = 0$$
$$36 - 84 + 48 = 0$$
$$0 = 0 \text{ True}$$

You can check the remaining examples on your own.

EXAMPLE 6 Solve: $n^2 - 11n + 24 = 0$

STEP 1 Factor the left side of the equation. Find two numbers that multiply to 24 and add to -11.

$$n^2 - 11n + 24 = 0$$

In order to get a negative sum and a positive product, both numbers must be negative. The numbers -3 and -8 will work.

$$(n - 3)(n - 8) = 0$$

STEP 2 Set each factor equal to 0 and solve for n.

$$n - 3 = 0 \qquad n - 8 = 0$$
$$n = 3 \qquad n = 8$$

The solution set is $\{3, 8\}$.

EXAMPLE 7 Solve: $b^2 + 7b - 60 = 0$

STEP 1 Factor the left side of the equation. Find two numbers that multiply to -60 and add to 7.

$$b^2 + 7b - 60 = 0$$

To get a negative product, the signs of the two numbers must be different. To get a positive sum, the number with the greater absolute value must be positive. The numbers 12 and -5 will work because $12 \times -5 = -60$ and $12 + -5 = 7$.

$$(b + 12)(b - 5) = 0$$

STEP 2 Set each factor equal to 0 and solve for b.

$$b + 12 = 0 \qquad b - 5 = 0$$
$$b = -12 \qquad b = 5$$

The solution set is $\{-12, 5\}$.

EXAMPLE 8   Solve: $m^2 - 2m - 63 = 0$

STEP 1   Factor the left side of the equation. Find two numbers that multiply to -63 and add to -2.

$$m^2 - 2m - 63 = 0$$

To get a negative product, the signs of the two numbers must be different. To get a negative sum, the number with the greater absolute value must be negative. The numbers -9 and 7 will work because $-9 \times 7 = -63$ and $-9 + 7 = -2$.

$$(m - 9)(m + 7) = 0$$

STEP 2   Set each factor equal to 0 and solve for m.

$$m - 9 = 0 \qquad m + 7 = 0$$
$$m = 9 \qquad m = -7$$

The solution set is $\{-7, 9\}$.

Watch for patterns in the positive and negative signs as you work. Some students find that memorizing the following patterns help them factor more quickly.

Pattern	Example
If all the terms in the trinomial are added, both factors show addition.	$x^2 + 4x + 3 = (x + 3)(x + 1)$
If only the bx term is subtracted, both factors show subtraction.	$x^2 - 4x + 3 = (x - 3)(x - 1)$
If only the c term is subtracted, the factors have $+$ and $-$ signs. The number with the greater absolute value is added.	$x^2 + 2x - 3 = (x + 3)(x - 1); 3 > 1$
If both bx and c are subtracted, the factors have $+$ and $-$ signs. The number with the greater absolute value is subtracted.	$x^2 - 2x - 3 = (x + 1)(x - 3); 3 > 1$

Solve by factoring.

6. $n^2 - 11n + 24 = 0$ $\qquad$ $x^2 + 2x - 48 = 0$ $\qquad$ $a^2 - 3a + 2 = 0$

7. $b^2 - 2b - 63 = 0$ $\qquad$ $m^2 + 11m + 28 = 0$ $\qquad$ $c^2 + 2c - 24 = 0$

8. $x^2 - 3x - 70 = 0$ $\qquad$ $a^2 + 15a + 50 = 0$ $\qquad$ $n^2 + 8n + 7 = 0$

9. $m^2 - 2m - 35 = 0$ $\qquad$ $b^2 + b - 72 = 0$ $\qquad$ $x^2 - 12x + 27 = 0$

10. $n^2 + 11n + 30 = 0$ $\qquad$ $a^2 - 2a - 8 = 0$ $\qquad$ $c^2 + 18c + 77 = 0$

11. $x^2 - 4x - 45 = 0$ $n^2 - n - 30 = 0$ $m^2 - 15m + 54 = 0$

12. $a^2 - 7a - 30 = 0$ $c^2 - 14c + 45 = 0$ $n^2 - 5n - 6 = 0$

13. $x^2 + x - 12 = 0$ $m^2 + 5m - 84 = 0$ $b^2 + 18b + 72 = 0$

14. $n^2 + 8n + 15 = 0$ $c^2 - 17c + 66 = 0$ $a^2 - 3a - 54 = 0$

In many quadratic expressions, the squared variable has a coefficient other than 1. One type has the form $ax^2 + bx$. To work with a quadratic in this form, factor out the greatest common factor. (**Note:** The form $ax^2 + bx$ is the same as $ax^2 + bx + c$ when $c = 0$.)

EXAMPLE 9   Solve: $5x^2 - 10x = 0$

STEP 1   Factor out the greatest common factor. The quadratic expression has two terms: $5x^2$ and $-10x$. The GCF for both terms is $5x$. Factor out $5x$.

$$5x^2 - 10x = 0$$
$$5x(x - 2) = 0$$

STEP 2   Set each factor equal to 0 and solve for x.

$$5x = 0 \qquad x - 2 = 0$$
$$x = 0 \qquad x = 2$$

The solution set is **{0, 2}**.

Solve by factoring.

15. $12n^2 + 48n = 0$ $3a^2 - 9a = 0$ $30x^2 + 54x = 0$

16. $7m^2 + 70m = 0$ $12x^2 - 8x = 0$ $16a^2 - 20a = 0$

17. $20b^2 - 50b = 0$ $45n^2 - 10n = 0$ $27x^2 + 18x = 0$

It is more challenging to solve quadratic expressions in the form $ax^2 + bx + c$ when $a \neq 1$. This kind of problem is a lot like solving a puzzle.

EXAMPLE 10 Solve: $2x^2 + 13x + 20 = 0$

STEP 1 Factor. The quadratic expression will be factored into two binomials. Based on your previous work with factoring, you should be able to draw some conclusions.

- The product of the two x terms in the binomials will be $2x^2$, so the factors must be $2x$ and x.

- The product of the two constant terms in the binomials will be 20. Possible factors include 1 and 20, 2 and 10, and 4 and 5.

You can now use trial and error to find the middle term. Write $2x$ and x as the first terms in the binomials, and then try various arrangements of the factors of 20 until one works.

$$(2x + \square)(x + \square) = 2x^2 + 13x + 20$$

You need an arrangement of factors that will result in $13x$ as the middle term. For each trial, find the sum of the Outer and Inner products.

- $(2x + 20)(x + 1)$ yields the middle term $2x + 20x = 22x$.
- $(2x + 1)(x + 20)$ yields the middle term $40x + x = 41x$.
- $(2x + 10)(x + 2)$ yields the middle term $4x + 10x = 14x$.
- $(2x + 2)(x + 10)$ yields the middle term $20x + 2x = 22x$.
- $(2x + 4)(x + 5)$ yields the middle term $10x + 4x = 14x$.
- $(2x + 5)(x + 4)$ yields the middle term $8x + 5x = 13x$.

Only the last arrangement of factors works: $2x^2 + 13x + 20 = 0$
$$(2x + 5)(x + 4) = 0$$

STEP 2 Set each factor equal to 0 and solve for x.

The solution set is $\left\{-\frac{5}{2}, -4\right\}$.

$$2x + 5 = 0 \qquad x + 4 = 0$$
$$2x = -5 \qquad x = -4$$
$$x = -\frac{5}{2}$$

As you gain more experience with factoring, you'll get faster at it.

EXAMPLE 11 Solve: $3a^2 - 7a - 6 = 0$

STEP 1 Factor.

- The factors of $3a^2$ must be $3a$ and a.

- The possible factors of -6 are -6 and 1, -1 and 6, -3 and 2, and -2 and 3.

Now use trial and error. You know that the signs of the factors will be different, but you need to determine which one is positive and which one is negative.

$$(3a \; \square)(a \; \square) = 3a^2 - 7a - 6$$

Study the pairs of factors for −6. One number in the pair will be multiplied by 3 and then added to the remaining factor. The result needs to be −7, the coefficient of the middle term.

Focus on the pair −3 and 2. If you multiply −3 by 3 and then add 2, you will get −7. Write the factorization: $(3a + 2)(a − 3) = 0$.

Always use FOIL to check that you have the right factors.
$(3a + 2)(a − 3) = 3a^2 − 9a + 2a − 6 = 3a^2 − 7a − 6$

STEP 2 Set each factor equal to 0 and solve for a.

$$\begin{array}{ll} 3a + 2 = 0 & a - 3 = 0 \\ 3a = -2 & a = 3 \\ a = -\frac{2}{3} & \end{array}$$

The solution set is $\left\{-\frac{2}{3}, 3\right\}$.

Sometimes the coefficient of the first term also needs to be factored. Then use what you learned about trial and error to help you find the solution set.

EXAMPLE 12 Factor: $9x^2 − 34x − 8$

STEP 1 Factor.

- The factors of $9x^2$ will be either $3x$ and $3x$ or $9x$ and x.

- The possible factors of −8 are −8 and 1, 8 and −1, −2 and 4, and 2 and −4.

$(3x \; \square)(3x \; \square)$ or $(9x \; \square)(x \; \square)$

Use trial and error. You need an arrangement of factors that will result in −34x as the middle term. As in Example 10, find the sum of the Outer and Inner products.

- $(3x − 8)(3x + 1)$ yields the middle term $3x − 24x = −21x$.

- $(3x + 8)(3x − 1)$ yields the middle term $−3x + 24x = 21x$.

- $(3x − 2)(3x + 4)$ yields the middle term $12x − 6x = 6x$.

- $(3x + 2)(3x − 4)$ yields the middle term $−12x + 6x = −6x$.

- $(9x − 8)(x + 1)$ yields the middle term $9x − 8x = x$.

- $(9x + 8)(x − 1)$ yields the middle term $−9x + 8x = −x$.

- $(9x − 2)(x + 4)$ yields the middle term $36x − 2x = 34x$.

- $(9x + 2)(x − 4)$ yields the middle term $−36x + 2x = −34x$.

The last arrangement of factors works.
$$9x^2 − 34x − 8 = 0$$
$$(9x + 2)(x − 4) = 0$$

STEP 2 Set each factor equal to 0 and solve for x.

$$\begin{array}{ll} 9x + 2 = 0 & x - 4 = 0 \\ 9x = -2 & x = 4 \\ x = -\frac{2}{9} & \end{array}$$

The solution set is $\left\{-\frac{2}{9}, 4\right\}$.

Solve by factoring.

18. $3x^2 - 28x + 49 = 0$ $2a^2 + 7a - 9 = 0$ $3n^2 - 16n + 5 = 0$

19. $7b^2 - 51b + 54 = 0$ $7m^2 + 2m - 9 = 0$ $2c^2 - 13c + 18 = 0$

20. $3a^2 + 19a - 110 = 0$ $2n^2 + 27n + 70 = 0$ $5b^2 + 63b + 36 = 0$

21. $5x^2 + 41x + 8 = 0$ $5m^2 - 27m + 10 = 0$ $3c^2 + 31c + 56 = 0$

22. $3n^2 - n - 2 = 0$ $4a^2 - 21a - 49 = 0$ $10x^2 + 9x - 1 = 0$

23. $10b^2 + 11b - 8 = 0$ $22x^2 + 37x + 6 = 0$ $25m^2 - 15m - 54 = 0$

24. $3n^2 + 11n - 20 = 0$ $11c^2 - 94c - 45 = 0$ $8a^2 + a - 9 = 0$

25. $21x^2 - 70x + 49 = 0$ $22b^2 - 35b - 50 = 0$ $6m^2 - 7m - 5 = 0$

26. $21a^2 - 48a + 27 = 0$ $8c^2 - 91c - 60 = 0$ $4n^2 + 33n - 27 = 0$

Special Factorizations

In this lesson, you will learn to

- Factor a difference of two squares.

- Factor the sum and difference of two cubes.

When working with polynomials, some patterns occur frequently. Recognizing those patterns can help you solve problems quickly.

A basic pattern occurs when you multiply $(a - b)(a + b)$. Notice that one factor, $a - b$, shows subtraction and the other, $a + b$, shows addition of the same values.

$$(a - b)(a + b)$$
$$= a^2 + ab - ab - b^2$$
$$= a^2 - b^2$$

The result $a^2 - b^2$ is called a **difference of squares.** Do you see why? Both a^2 and b^2 are perfect squares, and one is subtracted from the other.

Whenever you see a difference of two squares, you can immediately factor it using the pattern below.

Difference of Squares	$a^2 - b^2 = (a - b)(a + b)$

EXAMPLE 1 Factor: $x^2 - 25$

STEP 1 Recognize the pattern. The terms x^2 and 25 are both perfect squares ($25 = 5^2$). You know this is a difference of squares because one term is subtracted from the other.

STEP 2 Use the pattern to write the factorization.

$$a^2 - b^2 = (a - b)(a + b)$$
$$x^2 - 5^2 = (x - 5)(x + 5)$$

In the following example, the first term has a number coefficient. Because this coefficient is a perfect square, the entire expression is a difference of squares.

EXAMPLE 2 Factor: $16x^2 - 9$

STEP 1 Recognize the pattern. The term $16x^2$ is a perfect square because both 16 and x^2 are perfect squares: $16x^2 = (4x)^2$. The term 9 is also a perfect square: $9 = 3^2$.

STEP 2 Use the pattern to write the factorization.

$$a^2 - b^2 = (a - b)(a + b)$$
$$(4x)^2 - 3^2 = (4x - 3)(4x + 3)$$

Is there a pattern to a sum of squares? Can $a^2 + b^2$ be factored?

At first glance, you might think the correct factorization is $(a + b)(a + b)$. Try it: $(a + b)(a + b) = a^2 + 2ab + b^2$. It doesn't work. In fact, the expression $a^2 + b^2$ cannot be factored with real numbers for a and b.

You can use the difference of squares pattern to solve a quadratic equation.

EXAMPLE 3   Solve: $36x^2 = 25$

STEP 1   Rewrite with the quadratic expression equal to 0.

STEP 2   Write the equation as a difference of squares.

$$36x^2 = 25$$
$$36x^2 - 25 = 0$$
$$(6x)^2 - (5)^2 = 0$$
$$(6x - 5)(6x + 5) = 0$$

STEP 3   Set each factor equal to 0, and solve for x.

The solution is $\left\{-\frac{5}{6}, \frac{5}{6}\right\}$.

$$6x - 5 = 0 \qquad 6x + 5 = 0$$
$$6x = 5 \qquad\quad 6x = -5$$
$$x = \frac{5}{6} \qquad\quad x = -\frac{5}{6}$$

Sometimes you can apply a form more than once when factoring. For any problem, always make sure you have factored an expression completely.

EXAMPLE 4   Factor: $a^4 - 81$

STEP 1   This is a difference of squares because $a^4 = (a^2)^2$ and $81 = 9^2$.

STEP 2   Use the pattern to write the factorization.

STEP 3   Continue factoring. The factor $a^2 - 9$ is also a difference of squares. You can't factor the sum of squares, $a^2 + 9$.

$$a^4 - 81 = (a^2)^2 - (9)^2$$
$$= (a^2 - 9)(a^2 + 9)$$
$$= (a - 3)(a + 3)(a^2 + 9)$$

Two additional forms are helpful when factoring. When a binomial is squared, the result is a trinomial square. Study the forms below and see if you notice any patterns.

Forming Trinomial Squares

$$(a + b)^2 = (a + b)(a + b) = a^2 + 2ab + b^2$$
$$(a - b)^2 = (a - b)(a - b) = a^2 - 2ab + b^2$$

If you master these patterns, you'll be able to factor more quickly.

EXAMPLE 5   Factor: $x^2 + 8x + 16$

STEP 1   Recognize the pattern. The last term 16 is a perfect square whose positive root is 4. The middle coefficient is 8, which is double the root of the third term.

STEP 2   Apply the pattern.

$$x^2 + 8x + 16 = (x + 4)^2$$

You can check your work by performing the multiplication. Pay close attention to how the middle term is formed.

$$(x + 4)(x + 4) = x^2 + 4x + 4x + 16$$
$$= x^2 + 8x + 16$$

EXAMPLE 6 Factor: $4x^2 - 20x + 25$

STEP 1 Recognize the pattern. The first and the last terms are perfect squares. The square root of the first term is $2x$. The square root of the last term is 5. If you multiply these roots and double the result, you have the middle term, $20x$. Because the middle term is subtracted, you will use subtraction in the binomial.

STEP 2 Apply the pattern.

$$4x^2 - 20x + 25 = (2x - 5)^2$$

Check your work.

$$(2x - 5)(2x - 5) = 4x^2 - 10x - 10x + 25$$
$$= 4x^2 - 20x + 25$$

Of course, you could always factor these expressions using trial and error. Knowing the common forms, however, will help you see patterns and factor more easily.

Factor each expression completely.

1. $n^2 - 144$ $16x^2 - 1$ $81a^2 - 121$

2. $c^4 - 16$ $81n^4 - 1$ $x^2 - 12x + 36$

3. $x^2 - 16x + 64$ $25a^2 + 10a + 1$ $16x^2 - 24x + 9$

Solve by factoring.

4. $m^2 - 64 = 0$ $16a^2 - 1 = 0$ $49n^2 - 121 = 0$

5. $100c^2 - 49 = 0$ $x^2 - 24x + 144 = 0$ $9a^2 - 12a + 4 = 0$

6. $n^2 + 16n + 64 = 0$ $36b^2 - 60b + 25 = 0$ $64m^2 + 112m + 49 = 0$

Factoring by Grouping

In this lesson, you will learn to

- Factor by grouping the terms in a polynomial.

Sometimes the product of two binomials is a polynomial with four terms.

Look at the binomial factors $(7n + 3)(3n^2 + 5)$.

Multiply using FOIL, and then arrange the terms with variables in descending order.

$$(7n + 3)(3n^2 + 5) = 21n^3 + 35n + 9n^2 + 15$$
$$= 21n^3 + 9n^2 + 35n + 15$$

You now have a polynomial with four terms. You know this polynomial can be factored, but how would you go about factoring it? The answer involves grouping terms and applying the distributive property. Study the steps in the following example.

EXAMPLE 1   Factor: $21n^3 + 9n^2 + 35n + 15$

STEP 1   Create two groups of terms. Separate the groups with an addition symbol. It is important to keep track of that symbol.

$$\underline{21n^3 + 9n^2} \boxed{+} \underline{35n + 15}$$

STEP 2   Factor out the greatest common factor (GCF) from each group.

$$3n^2(7n + 3) \boxed{+} 5(7n + 3)$$

The GCF of $21n^3 + 9n^2$ is $3n^2$.
The GCF of $35n + 15$ is 5.

$$3n^2(7n + 3) + 5(7n + 3)$$

STEP 3   Put everything back together. You now have two terms with a common factor of $7n + 3$. Use the distributive property to factor out $7n + 3$ from both terms. The factored form of $21n^3 + 9n^2 + 35n + 15$ is $\mathbf{(7n + 3)(3n^2 + 5)}$.

$$\mathbf{(7n + 3)(3n^2 + 5)}$$

Make sure you understand where the factor $3n^2 + 5$ comes from. It comes from the remaining terms after $7n + 3$ is factored out.

Try another example. Remember, you want to group the terms so you have two expressions with a common binomial factor. When you factor out that binomial, the remaining terms form a second binomial factor.

EXAMPLE 2   Factor: $2x^3 - 10x^2 + x - 5$

STEP 1   Group the terms.

$$\underline{2x^3 - 10x^2} \boxed{+} \underline{x - 5}$$

STEP 2   Factor out the greatest common factor from each group. The GCF of the first group is $2x^2$.

$$2x^2(x - 5) \boxed{+} x - 5$$

There doesn't seem to be a GCF of the second group. Don't worry, the number 1 is a factor of anything.

$$2x^2(x - 5) + 1(x - 5)$$
$$(x - 5)(2x^2 + 1)$$

STEP 3 Factor out $x - 5$.

So far the groups of terms have been connected by addition. When the groups are connected by subtraction, an extra step is necessary.

EXAMPLE 3   Factor: $60a^3 + 10a^2 - 42a - 7$

STEP 1 Group the terms. Rewrite $-42a$ as $+ -42a$ so that an addition symbol is in the middle.

$$60a^3 + 10a^2 - 42a - 7$$
$$\underline{60a^3 + 10a^2} \boxed{+} \underline{-42a - 7}$$

STEP 2 Factor out the greatest common factor from each group.

$$10a^2(6a + 1) \boxed{+} -7(6a + 1)$$

The GCF of $60a^3 + 10a^2$ is $10a^2$.
The GCF of $-42a - 7$ is -7.

STEP 3 Put everything back together. Notice that the addition of a negative is written as subtraction. Factor out $6a + 1$ and write the two factors.

$$10a^2(6a + 1) - 7(6a + 1)$$
$$(6a + 1)(10a^2 - 7)$$

Some polynomials cannot be factored.

EXAMPLE 4   Factor: $m^3 + m^2 + 2m - 2$

STEP 1 Group the terms.

$$\underline{m^3 + m^2} \boxed{+} \underline{2m - 2}$$

STEP 2 Factor out the GCF.

$$m^2(m + 1) + 2(m - 1)$$

STEP 3 There is no common factor to factor out. $m + 1$ and $m - 1$ are different quantities. This polynomial cannot be factored.

Factoring is just one of the tools you can use when working with quadratic expressions and equations. Because factoring will not always work, it is helpful to use a strategy when you try to factor any expression.

Recommended Factoring Strategy

1. First look for a common factor of all the terms. If you find one, factor it out.

2. Then count the terms.

   - If there are 2 terms, look for a difference of squares.
   - If there are 3 terms, look for a trinomial square. If there is no trinomial square, use trial and error to find two binomial factors.
   - If there are 4 terms, try grouping the terms and finding a common GCF for both groups.

3. Always make sure each factor has been factored completely.

Factor each expression completely. Remember that some expressions cannot be factored.

1. $6x^3 + x^2 + 12x + 2$

$4a^3 + 28a^2 + 7a + 49$

$12m^3 - 9m^2 + 8x - 6$

2. $9b^3 - 8b^2 + 9b - 8$

$5x^3 + 15x^2 - 4x + 12$

$4n^3 + 3n^2 - 40n - 30$

3. $18x^3 + 9x^2 - 2x - 2$

$3a^3 + 7a^2 + 6a - 14$

$18n^3 + 9n^2 + 14n + 7$

4. $48a^3 + 30a^2 - 8a - 5$

$8x^3 + 16x^2 - 14x - 28$

$3n^3 + 4n^2 - 9n + 12$

5. $7x^3 + 5x^2 - 28x - 20$

$18a^3 - 2a^2 - 45a + 5$

$32m^3 + 20m^2 - 8m - 5$

6. $4a^3 - a^2 - 4a + 1$

$6x^3 + 2x^2 + 15x + 5$

$4b^3 + b^2 - 8b - 2$

7. $28x^3 - 36x^2 - 21x - 27$

$24b^3 - 12b^2 + 16b - 8$

$5m^3 + 25m^2 + 4m + 20$

8. $20n^3 + 12n^2 - 10n - 6$

$2x^3 - 2x^2 + 3x + 3$

$4a^3 - 8a^2 + 3a - 6$

9. $60m^3 - 24m^2 - 45m + 18$

$4n^3 + 3n^2 - 16n - 12$

$5b^3 - 2b^2 - 5b + 2$

Simplifying Rational Expressions

In this lesson, you will learn to

- Use factoring to write rational expressions in simplest form.

Working with more complicated expressions and equations requires tools to make your work more manageable. Two very helpful tools are factoring and canceling.

You can use **factoring** to write an expression as a product of terms. **Canceling** allows you to eliminate factors using the concept that a quantity divided by itself equals 1.

A **rational expression** is a quotient that contains a polynomial in both the numerator and the denominator. To simplify a rational expression, completely factor the numerator and the denominator. Then cancel when possible.

EXAMPLE 1 Simplify $\dfrac{x + 2}{3x^2 - 3x - 18}$. State which values for x must be excluded.

STEP 1 Completely factor the expression. First, factor out the common factor 3 from all the terms in the denominator.

$$\frac{x + 2}{3(x^2 - x - 6)}$$

Next, factor the trinomial.

$$\frac{x + 2}{3(x - 3)(x + 2)}$$

STEP 2 Cancel $x + 2$ in both the numerator and the denominator. You can do this because $\frac{x + 2}{x + 2} = 1$.

$$\frac{\cancel{x + 2}}{3(x - 3)\cancel{(x + 2)}} = \frac{1}{3(x - 3)}$$

So, $\dfrac{x + 2}{3x^2 - 3x - 18} = \dfrac{1}{3(x - 3)}$.

There is, however, one more step to this problem. When you work with rational expressions, certain values of the variable must be excluded because division by 0 is undefined. You must specify any restrictions on x.

STEP 3 Look back to the expression when it was completely factored: $\dfrac{x + 2}{3(x - 3)(x + 2)}$.

What values of x would make the denominator equal to 0? If any factor in the denominator equals 0, the entire denominator will equal 0. Set each factor equal to 0 and solve. These are the excluded values.

The excluded values for x are $\{-2, 3\}$.

$$x - 3 = 0 \qquad x + 2 = 0$$
$$x = 3 \qquad\quad x = -2$$

Do you see why you can't use only the simplified form to find the excluded values? Based on the simplified form, you would know to exclude 3, but not -2. Recognizing excluded values is a very important part of working with functions, radicals, and quadratic equations. Always use either the original expression or the expression in its fully factored form to determine the excluded values.

EXAMPLE 2 Simplify $\dfrac{2m + 2}{2m + 2}$. State which values for m must be excluded.

STEP 1 You can immediately see that the value of the expression is 1 because the numerator and the denominator are the same. Still, you should factor the expression.

$$\dfrac{\cancel{2(m + 1)}}{\cancel{2(m + 1)}} = 1$$

STEP 2 What values must be excluded? Set $m + 1$ equal to 0. The value -1 must be excluded.

$$m + 1 = 0$$
$$m = -1$$

The expression equals **1** and the excluded value is **{−1}**.

Simplify. Write excluded values in brackets. Remember, excluded values are only those values that make the denominator equal to 0.

1. $\dfrac{3x^2 + 14x - 5}{2x^2 + 12x + 10}$ $\dfrac{42a^2 + 6a}{12a^2 + 60a}$ $\dfrac{2x^2 - 19x + 9}{3x^2 - 23x - 36}$

2. $\dfrac{5n^2 + 35n}{3n^2 + 15n - 42}$ $\dfrac{2x^2 + 13x + 20}{5x^2 + 10x - 40}$ $\dfrac{3b^2 + 5b + 2}{9b^2 + 9b}$

3. $\dfrac{8a + 8}{6a^3 - 9a^2 - 15a}$ $\dfrac{6n^3 - 75n^2 + 150n}{3n^2 - 36n + 60}$ $\dfrac{4n^3 + 8n^2 - 12n}{3n^2 + 6n - 9}$

4. $\dfrac{8x + 40}{6x^3 + 36x^2 + 30x}$ $\dfrac{15b^3 - 87b^2 - 18b}{10b^3 - 80b^2 + 120b}$ $\dfrac{6x^2 - 15x - 54}{6x - 15}$

5. $\dfrac{18x - 12}{12x^2 - 48}$ $\dfrac{9m - 54}{6m^3 - 18m^2 - 108m}$ $\dfrac{8n^2 - 32n - 40}{4n - 20}$

6. $\dfrac{10a^3 + 14a^2 + 4a}{15a^2 + 21a + 6}$ $\dfrac{6n^2 - 9n + 3}{6n^3 - 12n^2 + 6n}$ $\dfrac{10x^2 - 40x + 30}{9x^2 - 39x + 36}$

More About Radicals

In this lesson, you will learn to

- Perform operations with radicals.
- Determine excluded values for radical expressions.
- Use conjugates to rationalize binomial denominators with radicals.

A **radical expression** is an expression that contains a radical. The part of the expression that is written inside the bracket is called the **radicand.**

You can add or subtract expressions that have like radicands.

EXAMPLE 1   Add: $2\sqrt{3} + 5\sqrt{3}$

The radicands are the same, so you can add the numbers in front of the radicands.

$$2\sqrt{3} + 5\sqrt{3} = 7\sqrt{3}$$

In addition and subtraction, a radicand acts like a variable. Ask yourself, "How would I solve this problem if I replaced $\sqrt{3}$ with x?" Instead of $2\sqrt{3} + 5\sqrt{3}$, think of it as $2x + 5x$. The answer is $7x$ or, in this example, $7\sqrt{3}$.

Sometimes the radicands do not start out alike. If you can, simplify one or both radicals so that both radicands are the same.

EXAMPLE 2   Subtract: $\sqrt{8} - 3\sqrt{2}$

STEP 1   First simplify $\sqrt{8}$. This process is explained on page 10.

$$\sqrt{8} = \sqrt{4 \cdot 2} = \sqrt{4} \cdot \sqrt{2} = 2\sqrt{2}$$

STEP 2   Rewrite the subtraction. The radicands are now alike. Subtract.

$$\sqrt{8} - 3\sqrt{2} = 2\sqrt{2} - 3\sqrt{2}$$
$$= -1\sqrt{2}$$
$$= -\sqrt{2}$$

You don't need to write the 1, just as you don't need to write 1 in front of an x term.

Remember, if the radicands are not the same, you can't combine the terms. The expression is already in its simplest form.

Add or subtract. Write the answer in simplest form.

1. $3\sqrt{7} + 5\sqrt{7}$ $4\sqrt{6} - \sqrt{6}$ $-3\sqrt{10} + 4\sqrt{10}$

2. $3\sqrt{8} - 4\sqrt{8}$ $10\sqrt{20} + 7\sqrt{20}$ $-5\sqrt{18} + 4\sqrt{18}$

3. $3\sqrt{50} - 2\sqrt{2}$ $\qquad$ $-\sqrt{63} + 3\sqrt{7}$ $\qquad$ $-4\sqrt{24} - 5\sqrt{6}$

4. $-2\sqrt{8} + 4\sqrt{200}$ $\qquad$ $5\sqrt{40} - 4\sqrt{90}$ $\qquad$ $3\sqrt{125} + 2\sqrt{45}$

5. $3\sqrt{32} + 2\sqrt{8}$ $\qquad$ $-3\sqrt{12} - 5\sqrt{48}$ $\qquad$ $-5\sqrt{200} + 4\sqrt{8}$

As you move on to multiplying and dividing radicals, you will begin to use variables. You need to be comfortable with radicals because you will need to use them to solve quadratic equations. You can review the basic steps for multiplying and dividing radicals on pages 9–12.

EXAMPLE 3 Multiply: $\sqrt{15x^3} \cdot \sqrt{6x}$

STEP 1 Multiply the radicands.

$$\sqrt{15x^3 \cdot 6x} = \sqrt{90x^4}$$

STEP 2 Simplify. Do you see why $\sqrt{x^4} = x^2$?

$$\sqrt{9 \cdot 10 \cdot x^2 \cdot x^2} = 3x^2\sqrt{10}$$

A binomial can contain a radical. Use the distributive property to find the product.

EXAMPLE 4 Multiply: $\sqrt{2b}(2 + 3\sqrt{3b})$

The binomial $2 + 3\sqrt{3b}$ is multiplied by the single term $\sqrt{2b}$.

STEP 1 Use the distributive property.

$$\sqrt{2b}(2 + 3\sqrt{3b}) = (\sqrt{2b} \cdot 2) + (\sqrt{2b} \cdot 3\sqrt{3b})$$

STEP 2 Simplify.

$$= 2\sqrt{2b} + 3\sqrt{6b^2}$$
$$= 2\sqrt{2b} + 3b\sqrt{6}$$

You also can multiply two binomials containing radical expressions. Apply FOIL. Then simplify each term. First, try one without variables.

EXAMPLE 5 Multiply: $(-5 + \sqrt{2})(-4 - 5\sqrt{2})$

STEP 1 Use FOIL.

First: $-5 \cdot -4 = 20$ $\qquad$ Outer: $-5 \cdot -5\sqrt{2} = 25\sqrt{2}$

Inner: $\sqrt{2} \cdot -4 = -4\sqrt{2}$ $\qquad$ Last: $\sqrt{2} \cdot -5\sqrt{2} = -5 \cdot 2 = -10$

STEP 2 Combine like terms.

$$(20 - 10) + (25\sqrt{2} - 4\sqrt{2}) = 10 + 21\sqrt{2}$$

The next example contains a variable, but the process remains the same.

EXAMPLE 6 Multiply: $(-1 + 3\sqrt{5x})(-2 + \sqrt{5x})$

STEP 1 Use FOIL.
First: $-1 \cdot -2 = 2$ Outer: $-1 \cdot \sqrt{5x} = -\sqrt{5x}$
Inner: $3\sqrt{5x} \cdot -2 = -6\sqrt{5x}$ Last: $3\sqrt{5x} \cdot \sqrt{5x} = 3 \cdot 5x = 15x$

STEP 2 Combine like terms.
$$2 + (-\sqrt{5x} - 6\sqrt{5x}) + 15x = 2 - 7\sqrt{5x} + 15x$$

In simplest terms, the product is $2 - 7\sqrt{5x} + 15x$.

Multiply and simplify.

6. $\sqrt{3a^2} \cdot \sqrt{4a}$ $-5\sqrt{6x^3} \cdot -2\sqrt{6x^2}$ $5\sqrt{2m^2} \cdot 5\sqrt{2m^2}$

7. $\sqrt{6b}\left(4\sqrt{3} + 2\right)$ $-4\sqrt{3}\left(\sqrt{3} + \sqrt{10x}\right)$ $\sqrt{6n}\left(\sqrt{2n} + 5\right)$

8. $\sqrt{15}\left(\sqrt{5x} + 2\sqrt{6}\right)$ $-4\sqrt{10n}(3 + \sqrt{10n})$ $\sqrt{5}\left(-4\sqrt{5b} + 5b^2\right)$

9. $\left(-2 - 2\sqrt{3}\right)\left(-5 + \sqrt{3}\right)$ $\left(5 + \sqrt{2}\right)\left(-5 + \sqrt{2}\right)$ $\left(\sqrt{3} + \sqrt{5}\right)\left(\sqrt{3} + 3\sqrt{5}\right)$

10. $\left(\sqrt{3x} - 3\right)\left(\sqrt{3x} + 5\right)$ $\left(3\sqrt{5} + 2\sqrt{2a}\right)\left(\sqrt{5} - 2\sqrt{2a}\right)$ $\left(\sqrt{5m} - \sqrt{2m}\right)\left(\sqrt{5m} + \sqrt{2m}\right)$

As you move on to division, remember that an expression is not simplified if there is a radical in the denominator. To divide radicals, you need to rationalize the denominator (see page 11), and then combine like terms and cancel if possible.

EXAMPLE 7 Simplify: $\frac{2}{\sqrt{5}}$

Multiply the fraction by $\frac{\sqrt{5}}{\sqrt{5}}$ to rationalize the denominator.
$$\frac{2}{\sqrt{5}} \cdot \frac{\sqrt{5}}{\sqrt{5}} = \frac{2\sqrt{5}}{5}$$

Multiplying by 1 will never change the value of a fraction.

The process is basically the same when variables are involved. In the next example, the numerator of the expression is a binomial.

EXAMPLE 8 Simplify: $\dfrac{\sqrt{2x^2} - 3}{\sqrt{7x^2}}$

STEP 1 Simplify each term.

$$\frac{\sqrt{2x^2} - 3}{\sqrt{7x^2}} = \frac{x\sqrt{2} - 3}{x\sqrt{7}}$$

STEP 2 Multiply the entire fraction by $\dfrac{\sqrt{7}}{\sqrt{7}}$ to rationalize the denominator.

$$\frac{x\sqrt{2} - 3}{x\sqrt{7}} \cdot \frac{\sqrt{7}}{\sqrt{7}} = \frac{x\sqrt{14} - 3\sqrt{7}}{7x}$$

The numerator has two terms, so use the distributive property to multiply both terms by $\sqrt{7}$.

When an expression has a binomial in the denominator, you can rationalize the denominator by multiplying both parts of the fraction by the **conjugate** of the denominator. Binomials are conjugates when their only difference is the sign of one of the terms.

Conjugates	$a\sqrt{b} + c\sqrt{d}$ and $a\sqrt{b} - c\sqrt{d}$

EXAMPLE 9 Simplify: $\dfrac{2}{2 - \sqrt{n}}$

Eliminate the radical in the denominator by multiplying the numerator and denominator by the conjugate of the denominator. This is the same as multiplying by 1.

$$\frac{2}{2 - \sqrt{n}} \cdot \frac{2 + \sqrt{n}}{2 + \sqrt{n}} = \frac{4 + 2\sqrt{n}}{4 - n}$$

Why does it work?

Work through multipying by the conjugate slowly using FOIL.

$(2 - \sqrt{n})(2 + \sqrt{n})$ First: $2 \cdot 2 = 4$ Outer: $2 \cdot \sqrt{n} = 2\sqrt{n}$

Inner: $-\sqrt{n} \cdot 2 = -2\sqrt{n}$ Last: $-\sqrt{n} \cdot \sqrt{n} = -n$

Combine the results: $4 + 2\sqrt{n} - 2\sqrt{n} - n = 4 - n$

The two terms in the middle combine to equal 0, so you are left with $4 - n$.

Does this remind you of anything? Think about the difference of squares. When you multiply $(x - 2)(x + 2)$, you get $x^2 - 2x + 2x - 4$. The two middle terms equal 0, and you are left with $x^2 - 4$.

Now that you see how conjugates work, you can multiply them without using FOIL. For example, to find $(\sqrt{x} + 3)(\sqrt{x} - 3)$, think: $\sqrt{x} \cdot \sqrt{x} = x$ and $3 \cdot -3 = -9$. The *product must be $x - 9$*. Recognizing this pattern can save you a lot of time.

Simplify.

11. $\dfrac{3\sqrt{4a}}{4\sqrt{5a}}$ $\dfrac{5x}{5\sqrt{2x^2}}$ $\dfrac{4\sqrt{3m}}{2\sqrt{2m^4}}$

12. $\dfrac{\sqrt{3x^2}}{2\sqrt{5x^2}}$ $\dfrac{4}{\sqrt{2n^4}}$ $\dfrac{3\sqrt{9b^2}}{\sqrt{15b}}$

13. $\dfrac{-1-\sqrt{2x}}{5\sqrt{5x^4}}$ $\dfrac{4a^2-\sqrt{2a^2}}{3\sqrt{12a^2}}$ $\dfrac{2\sqrt{3m^3}-4\sqrt{3m}}{\sqrt{3m^4}}$

14. $\dfrac{-1-3\sqrt{2n^2}}{5\sqrt{2n^2}}$ $\dfrac{3x+3\sqrt{3x^3}}{\sqrt{12x^4}}$ $\dfrac{-2-4\sqrt{b^2}}{\sqrt{18b^2}}$

15. $\dfrac{\sqrt{5a}}{5+\sqrt{5a^3}}$ $\dfrac{2}{4-\sqrt{2x}}$ $\dfrac{5}{3+\sqrt{2m}}$

16. $\dfrac{2}{-5+\sqrt{b^3}}$ $\dfrac{\sqrt{6n^2}}{2\sqrt{3n}+4}$ $\dfrac{4\sqrt{3a}}{6+6\sqrt{5a}}$

17. $-\dfrac{1}{5x^2+\sqrt{2x}}$ $\dfrac{5}{-2-5\sqrt{3a^3}}$ $\dfrac{4}{4m^2+\sqrt{2m^4}}$

Completing the Square

In this lesson, you will learn to

- Solve quadratic equations using perfect squares.
- Solve quadratic equations by completing the square.

A quadratic equation can have one, two, or no real-number roots. The simplest quadratic equation takes the form $x^2 = k$, where k is a constant.

EXAMPLE 1 Solve for x: $x^2 = 25$

Take the square root of both sides of the equation.

The solution set is {5, −5}.

$$x^2 = 25$$
$$x = \pm\sqrt{25}$$
$$x = \pm 5$$

> **Math Talk**
> The symbol $\pm$ means "plus or minus."

The next problem adds one more step. Make sure you understand why to solve the equation twice in Step 2.

EXAMPLE 2 Solve for a: $(a + 4)^2 = 36$

STEP 1 Take the square root of both sides of the equation. Remember, a positive number has two real roots—positive and negative.

$$(a + 4)^2 = 36$$
$$a + 4 = \pm\sqrt{36}$$
$$a + 4 = \pm 6$$

STEP 2 Solve the equation twice—once using +6 and once using −6.

$$a + 4 = 6 \qquad a + 4 = -6$$
$$a = 2 \qquad\qquad a = -10$$

The solution set is {2, −10}.

In these examples, x^2 and $(a + 4)^2$ are examples of **perfect squares.** A perfect square is simply an expression that is squared. The expression $(a + 4)^2$ is a binomial square.

You've already practiced factoring quadratic expressions. For certain trinomials, the factorization is a binomial square. You can use binomial squares to solve equations.

EXAMPLE 3 Solve for n: $n^2 + 10n + 25 = 16$

STEP 1 Factor the trinomial. The result is a binomial square.

STEP 2 Take the square root of both sides.

$$n^2 + 10n + 25 = 16$$
$$(n + 5)^2 = 16$$
$$n + 5 = \pm\sqrt{16}$$
$$n + 5 = \pm 4$$

STEP 3 Solve the equation using +4 and −4.

$$n + 5 = 4 \qquad n + 5 = -4$$
$$n = -1 \qquad\qquad n = -9$$

The solution set is {−1, −9}.

Study the following example to learn what to do when the answer to a quadratic equation is an irrational number.

EXAMPLE 4 Solve for x: $x^2 - 6x + 9 = 5$

STEP 1 Write the trinomial as a binomial square.

$$x^2 - 6x + 9 = 5$$
$$(x - 3)^2 = 5$$

STEP 2 Take the square root of both sides.

$$x - 3 = \pm\sqrt{5}$$
$$x = 3 \pm \sqrt{5}$$

STEP 3 The square root of 5 is an irrational number. Write the solution set using radicals.

$$\{3 + \sqrt{5}, 3 - \sqrt{5}\}$$

Using a calculator, you can write the solution with decimals. These solutions are irrational because the value of the radical continues without ending or repeating. Here the values are rounded to five decimal places.

$$3 + \sqrt{5} \approx 5.23607$$
$$3 - \sqrt{5} \approx 0.76393$$

Unless you are asked to use decimals, always express a solution set using exact values by writing the expressions with radicals.

All the examples you have seen so far have a positive number on the right side of the equation. What happens if that number is negative?

How would you solve the equation $x^2 = -9$? Ask yourself, "What number multiplied by itself equals -9?" There isn't one. If you square a positive number, the result is positive. If you square a negative number, the result is positive. You can't square a number and get a negative, real-number result.

When a perfect square is set equal to a negative number, there is no real-number solution. You could express the solution as an imaginary number, however, the topic of imaginary numbers is beyond the scope of this book.

EXAMPLE 5 Solve for m: $(m + 3)^2 = -2$

STEP 1 Take the square root of both sides.

$$(m + 3)^2 = -2$$
$$m + 3 = \pm\sqrt{-2}$$

Stop! $\sqrt{-2}$ is not a real number. There is **no real-number solution** to this problem.

Solve. Write *no solution* if the equation has no real number solution.

1. $(a + 5)^2 = 49$ $\qquad\qquad (x + 1)^2 = 81 \qquad\qquad (n + 3)^2 = -9$

2. $m^2 - 8m + 16 = 1$ $n^2 + 6n + 9 = -1$ $x^2 - 16x + 64 = 25$

3. $a^2 + 18a + 81 = 0$ $4m^2 + 12m + 9 = 4$ $9x^2 - 6x + 1 = 49$

4. $25b^2 + 20b + 4 = -4$ $4x^2 - 20x + 25 = 1$ $36n^2 - 12n + 1 = 36$

Examples 1 through 4 in this lesson show that it is always possible to solve a quadratic equation when there is a perfect square equal to a positive number. A quadratic equation can be transformed into this form by a process called **completing the square.**

EXAMPLE 6   Solve for x: $x^2 - 4x - 11 = 10$

STEP 1   Isolate the terms that contain x on the left side of the equation. Add 11 to both sides.

$$x^2 - 4x - 11 = 10$$
$$x^2 - 4x = 21$$

STEP 2   You need to turn the left side of the equation into a trinomial square. Take half of the middle term, 4, and square it: $\left(\frac{4}{2}\right)^2 = 4$. Add this amount to both sides of the equation.

$$x^2 - 4x + 4 = 21 + 4$$
$$x^2 - 4x + 4 = 25$$

STEP 3   Write the left side as a perfect square.

$$(x - 2)^2 = 25$$

STEP 4   Find the square root of both sides.

$$x - 2 = \pm\sqrt{25}$$
$$x - 2 = \pm 5$$

STEP 5   Write and solve two equations.

$$x - 2 = 5 \qquad x - 2 = -5$$
$$x = 7 \qquad\quad x = -3$$

The solution set is $\{7, -3\}$.

Look over Example 6. In Step 2, the addition property of equality allows for the addition of the same quantity to both sides of the equation. The amount of 4 was chosen so that the resulting trinomial on the left side of the equation could be represented by a perfect square.

Follow these steps to create a trinomial square.
For $x^2 + bx +$ _____

- Take half of b and square it: $\left(\frac{b}{2}\right)^2$
- Write the trinomial $x^2 + bx + \left(\frac{b}{2}\right)^2$, which equals $\left(x + \frac{b}{2}\right)^2$.

Forming a trinomial square can often introduce fractions into the equation. Study the next example to see how to work with fractions.

EXAMPLE 7   Solve for x: $x^2 + 3x - 42 = 12$

STEP 1   Isolate the terms with x on the left side of the equation. Leave room to complete the square.

$$x^2 + 3x - 42 = 12$$
$$x^2 + 3x \underline{\hspace{2cm}} = 54$$

STEP 2   The coefficient of the middle term is 3. Take half of 3 and square it. Then add $\left(\frac{3}{2}\right)^2$ to both sides of the equation.

$$x^2 + 3x + \left(\frac{3}{2}\right)^2 = 54 + \left(\frac{3}{2}\right)^2$$
$$x^2 + 3x + \frac{9}{4} = \frac{216}{4} + \frac{9}{4}$$
$$x^2 + 3x + \frac{9}{4} = \frac{225}{4}$$

STEP 3   Write the left side as a perfect square.

$$\left(x + \frac{3}{2}\right)^2 = \frac{225}{4}$$

STEP 4   Find the square root of both sides.

$$x + \frac{3}{2} = \pm\sqrt{\frac{225}{4}}$$

STEP 5   Solve in terms of x.

$$x = -\frac{3}{2} \pm \sqrt{\frac{225}{4}}$$

STEP 6   Write the solutions in simplest form.

Simplify the radical: $\sqrt{\frac{225}{4}} = \frac{\sqrt{225}}{\sqrt{4}} = \frac{\sqrt{15 \cdot 15}}{\sqrt{2 \cdot 2}} = \frac{15}{2}$

$$x = -\frac{3}{2} \pm \frac{15}{2}$$

Simplify the solutions: $x = -\frac{3}{2} + \frac{15}{2}$ $\qquad$ $x = -\frac{3}{2} - \frac{15}{2}$

$$= \frac{12}{2} \qquad\qquad\qquad = -\frac{18}{2}$$
$$= 6 \qquad\qquad\qquad = -9$$

The $\pm$ symbol tells you to solve the equation twice—once by adding $\frac{15}{2}$ and once by subtracting $\frac{15}{2}$.

The solution set is {6, −9}.

Wait a minute! Take another look at that original equation. What would happen if you subtracted 12 from both sides? You could factor it! And it would take a lot fewer steps!

$$x^2 + 3x - 42 = 12$$
$$x^2 + 3x - 54 = 0$$
$$(x + 9)(x - 6) = 0$$
$$x + 9 = 0 \qquad x - 6 = 0$$
$$x = -9 \qquad\quad x = 6$$

The values of x must be **−9 and 6**.

You may be wondering why you need to learn how to complete the square. After all, even though completing the square works, the process has a lot of steps, and factoring is much easier. If that's what you are thinking, then you are correct! Whenever possible, you should solve quadratic equations by factoring.

Sometimes, however, expressions contain fractions or radicals, making it very difficult or impossible to guess the factors. That is why you need to learn a way that will always work. Don't give up on completing the square!

Completing the square works only when the a term equals 1 in the trinomial form $ax^2 + bx + c$. You can use division, or multiplication by $\frac{1}{a}$, to make this coefficient equal to 1.

EXAMPLE 8   Solve for x: $3x^2 - 10x + 5 = 0$

STEP 1   Divide every term by 3, or multiply every term by $\frac{1}{3}$.

$$\frac{3}{3}x^2 - \frac{10}{3}x + \frac{5}{3} = \frac{0}{3}$$

(Note: It's okay to divide 0 by a number. In fact, 0 divided by any number equals 0.)

$$x^2 - \frac{10}{3}x + \frac{5}{3} = 0$$

STEP 2   Subtract $\frac{5}{3}$ from both sides. Leave room to complete the square.

$$x^2 - \frac{10}{3}x + \underline{\hspace{2cm}} = -\frac{5}{3}$$

STEP 3   Complete the square.

Find half of $-\frac{10}{3}$:

$$-\frac{10}{3} \cdot \frac{1}{2} = -\frac{10}{6} = -\frac{5}{3}$$

Square the result:

$$\left(-\frac{5}{3}\right)^2 = \frac{25}{9}$$

Add this term to both sides of the equation.

$$x^2 - \frac{10}{3}x + \frac{25}{9} = -\frac{5}{3} + \frac{25}{9}$$

Remember to use common denominators when you add fractions: $-\frac{5}{3} + \frac{25}{9} = -\frac{15}{9} + \frac{25}{9} = \frac{10}{9}$.

$$x^2 - \frac{10}{3}x + \frac{25}{9} = \frac{10}{9}$$

STEP 4   Write the left side as a perfect square.

$$\left(x - \frac{5}{3}\right)^2 = \frac{10}{9}$$

STEP 5   Find the square root of both sides.

$$x - \frac{5}{3} = \pm\sqrt{\frac{10}{9}}$$

Make sure to rationalize the denominator.

$$x - \frac{5}{3} = \pm\frac{\sqrt{10}}{3}$$

$$x = \frac{5}{3} \pm \frac{\sqrt{10}}{3}$$

The solution set is $\left\{\dfrac{5 + \sqrt{10}}{3}, \dfrac{5 - \sqrt{10}}{3}\right\}$.

$$x = \frac{5 \pm \sqrt{10}}{3}$$

The problems on the next page can be solved by completing the square. Some also can be solved by factoring. Remember, before you can solve a quadratic equation by factoring, you must rewrite the equation so that the trinomial equals 0.

As you look over your work, notice that when the roots of a quadratic equation are integers—positive and negative whole numbers—the equation could have been solved by factoring. If the answer includes radicals, however, the roots are irrational and you could not have solved the problem using factoring.

Solve by completing the square or by factoring. Write irrational roots in simplest radical form.

5. $a^2 - 10a + 2 = -5$

$n^2 - 4n - 23 = -3$

$x^2 - 2x - 5 = 4$

6. $x^2 + 5x + 10 = 6$

$b^2 - 8b + 8 = 3$

$m^2 - 6m - 8 = -2$

7. $4n^2 - 8n - 18 = 3$

$2x^2 - 4x - 19 = 3$

$5a^2 - 10a - 19 = -4$

8. $2x^2 - 8x - 17 = -5$

$4b^2 + 8b - 9 = -4$

$7n^2 - 14n - 62 = -6$

9. $8a^2 - 7a - 57 = -6$

$5x^2 + 6x - 51 = 5$

$6m^2 - m - 36 = 4$

10. $3x^2 + 16x - 21 = 6$

$6n^2 + 11n - 2 = -6$

$8b^2 + 4b - 5 = 5$

11. $3a^2 + 16a + 7 = -5$

$5n^2 - 14n - 5 = 2$

$x^2 + 3x - 24 = 3$

12. $3n^2 - 16n + 7 = 4$

$8x^2 + 8x - 31 = 7$

$3m^2 - m - 25 = 2$

The Quadratic Formula

In this lesson, you will learn to

- Solve quadratic equations using the quadratic formula.
- Use the discriminant to determine the number of solutions.

Quadratic equations take the form $ax^2 + bx + c = 0$, where $a \neq 0$. You have seen how to solve for x by factoring and by completing the square. To complete the square, you used a sequence of steps. Mathematicians have taken those same steps and applied them directly to the equation $ax^2 + bx + c = 0$ to create a very useful formula. Carefully follow the steps below.

First, divide each term by a. You can also think of it as multiplying each term by $\frac{1}{a}$.

$$ax^2 + bx + c = 0$$
$$x^2 + \frac{b}{a}x + \frac{c}{a} = 0$$

Add $-\frac{c}{a}$ to both sides to isolate the x terms on the left side of the equation.

$$x^2 + \frac{b}{a}x \quad = -\frac{c}{a}$$

Now complete the square. Half of $\frac{b}{a}$ is $\frac{b}{2a}$. The square of $\frac{b}{2a}$ is $\frac{b^2}{4a^2}$. Add this to both sides.

$$x^2 + \frac{b}{a}x + \frac{b^2}{4a^2} = -\frac{c}{a} + \frac{b^2}{4a^2}$$

Write the left side of the equation as a perfect square.

$$\left(x + \frac{b}{2a}\right)^2 = -\frac{c}{a} + \frac{b^2}{4a^2}$$

Add the terms on the right. The LCD is $4a^2$.

$$\left(x + \frac{b}{2a}\right)^2 = \frac{b^2 - 4ac}{4a^2}$$

$$-\frac{c}{a} + \frac{b^2}{4a^2} = -\left(\frac{4a}{4a}\right)\left(\frac{c}{a}\right) + \frac{b^2}{4a^2} = -\frac{4ac}{4a^2} + \frac{b^2}{4a^2} = \frac{b^2 - 4ac}{4a^2}$$

Find the square root of both sides.

$$x + \frac{b}{2a} = \pm\sqrt{\frac{b^2 - 4ac}{4a^2}}$$

Rationalize the denominator.

$$x + \frac{b}{2a} = \frac{\sqrt{b^2 - 4ac}}{2a}$$

Next, isolate x by subtracting $\frac{b}{2a}$ from both sides.

$$x = -\frac{b}{2a} \pm \frac{\sqrt{b^2 - 4ac}}{2a}$$

Finally, put it all together. This is the **quadratic formula.**

$$x = \frac{-b \pm \sqrt{b^2 - 4ac}}{2a}$$

Do you recognize the steps for completing the square? By performing the steps with variables instead of numbers, you have developed a formula that can be used as a shortcut to solve a quadratic equation.

The quadratic formula is such an important and useful tool that you need to be sure to master it.

The Quadratic Formula

If $ax^2 + bx + c = 0$ and $a \neq 0$, then $x = \dfrac{-b \pm \sqrt{b^2 - 4ac}}{2a}$.

EXAMPLE 1 Solve for x: $2x^2 + 9x - 3 = 0$

STEP 1 The equation is in the form $ax^2 + bx + c = 0$.

Find the values of a, b, and c.

$a = 2$
$b = 9$
$c = -3$

STEP 2 Substitute these values into the quadratic formula and simplify.

$$x = \frac{-b \pm \sqrt{b^2 - 4ac}}{2a}$$

$$x = \frac{-9 \pm \sqrt{9^2 - 4(2)(-3)}}{2(2)}$$

$$x = \frac{-9 \pm \sqrt{81 + 24}}{4}$$

The solution set is $\left\{ \dfrac{-9 + \sqrt{105}}{4}, \dfrac{-9 - \sqrt{105}}{4} \right\}$.

$$x = \frac{-9 \pm \sqrt{105}}{4}$$

Remember that the equation must be in the form $ax^2 + bx + c = 0$ before you can solve it.

EXAMPLE 2 Solve for x: $5x^2 - 8x = 13$

STEP 1 Write the equation in the form $ax^2 + bx + c = 0$, and find the values of a, b, and c.

$5x^2 - 8x = 13$
$5x^2 - 8x - 13 = 0$

$a = 5$
$b = -8$
$c = -13$

STEP 2 Substitute these values into the quadratic formula and simplify.

$$x = \frac{-b \pm \sqrt{b^2 - 4ac}}{2a}$$

Since the value of b is -8, the value of $-b$ is 8.

$$x = \frac{-(-8) \pm \sqrt{(-8)^2 - 4(5)(-13)}}{2(5)}$$

 Caution: $\sqrt{64 + 260}$ does NOT equal $\sqrt{64} + \sqrt{260}$. You must add first, and then find the square root.

$$x = \frac{8 \pm \sqrt{64 + 260}}{10}$$

$$x = \frac{8 \pm \sqrt{324}}{10}$$

You're almost finished! You need to break this into two expressions: one adding 18 and one subtracting 18.

$$x = \frac{8 \pm 18}{10}$$

STEP 3 Write and simplify the two solutions.

$$x = \frac{8 + 18}{10} \qquad x = \frac{8 - 18}{10}$$

The solution set is $\left\{ \dfrac{13}{5}, -1 \right\}$.

$$x = \frac{26}{10} = \frac{13}{5} \qquad x = \frac{-10}{10} = -1$$

Solve each equation for x using the quadratic formula.

1. $2x^2 - 7x - 72 = 0$ $x^2 - 5 = -4x$ $4x^2 + 7x = 36$

2. $3x^2 - 5x - 42 = 0$ $5x^2 - 4x = 3$ $8x^2 - 20 = 0$

     (Hint: $b = 0$)

3. $x^2 + 8x = 12$ $x^2 = 3x + 19$ $x^2 + 7x = -12$

4. $3x^2 - 8 = 10x$ $2x^2 - 1 = 0$ $x^2 - x = 11$

5. $3x^2 = 16$ $16x^2 + 25 = -40x$ $10x^2 - 2x = 2$

6. $x^2 - 4 = 10x$ $7x^2 - 10x = 1$ $4x^2 - 20x + 25 = 0$

7. $2x^2 = 7$ $4x^2 - 3x = 45$ $3x^2 = 17 + 6x$

You know that quadratic equations can have two, one, or zero roots. When an equation has zero roots, it has no real number solutions. You may be wondering whether there is any way you could tell what to expect just by looking at an equation.

One small part of the quadratic formula is called the **discriminant.** The discriminant is the expression in the radicand, $b^2 - 4ac$. You can substitute the values of a, b, and c from the equation and evaluate the discriminant.

- If the discriminant equals a positive number, there are two roots.
- If the discriminant equals 0, there is one root.
- If the discriminant equals a negative number, there are no real roots.

EXAMPLE 3 For each equation below, use the discriminant to determine the number of roots in the solution set.

$$6x^2 + 4x - 1 = 0 \qquad 9x^2 - 11x + 8 = 0 \qquad x^2 - 6x + 9 = 0$$

STEP 1 Identify a, b, and c for each equation.

$a = 6$	$a = 9$	$a = 1$
$b = 4$	$b = -11$	$b = -6$
$c = -1$	$c = 8$	$c = 9$

STEP 2 Solve for the discriminant.

$b^2 - 4ac$	$b^2 - 4ac$	$b^2 - 4ac$
$= (4)^2 - 4(6)(-1)$	$= (-11)^2 - 4(9)(8)$	$= (-6)^2 - 4(1)(9)$
$= 16 + 24$	$= 121 - 288$	$= 36 - 36$
$= 40$	$= -167$	$= 0$

STEP 3 Interpret the results.

40 is positive, so there are two roots. | −167 is negative, so there is no solution. | The discriminant equals 0, so there is one root.

Use the discriminant to determine the number of roots each equation has. Write *one*, *two*, or *no solution*. Then find the roots using any method.

8. $5x^2 + 6x = 27$ 　　　　 $9x^2 + 1 = 2x$ 　　　　 $2x^2 + 8 = 8x$

9. $5x^2 - 10x + 9 = 0$ 　　　　 $2x^2 + 7 = 10x$ 　　　　 $4x^2 + 3x - 52 = 0$

10. $9x^2 - 42x + 49 = 0$ 　　　　 $4x^2 - 5x + 10 = 0$ 　　　　 $5x^2 = 4x + 8$

Quadratic Equations and Word Problems

In this lesson, you will learn to

- Solve word problems using quadratic equations.

Quadratic equations create an interesting problem in practical applications. Most quadratic equations yield two answers, but how can a word problem have two answers? If you use a quadratic equation to solve a word problem, you must go back and check to see which answer fits the situation. An answer that doesn't fit is called an **extraneous solution.**

EXAMPLE 1   Find two consecutive negative numbers that multiply to 72.

STEP 1   Let x represent the first number and $x + 1$ represent the second number.

STEP 2   The product of the numbers is 72.

$$x(x + 1) = 72$$

STEP 3   Write the equation in quadratic form.

$$x^2 + x = 72$$
$$x^2 + x - 72 = 0$$
$$(x + 9)(x - 8) = 0$$

STEP 4   Solve. The factoring method is shown, but you could complete the square or use the quadratic formula to find the same answers.

$$x + 9 = 0 \qquad x - 8 = 0$$
$$x = -9 \qquad x = 8$$

STEP 5   Check the answers against the conditions in the problem. The problem states that the numbers are negative. Only −9 fulfills the requirement of the problem. The value 8 is an extraneous solution.

STEP 6   Apply the answer. If $x = -9$, then $x + 1 = -8$. The two numbers are **−9 and −8.**

Many science formulas use quadratic equations. For example, when a projectile is thrown or launched upward, you can determine its approximate height from the ground after a certain number of seconds using the formula: $h = -5t^2 + vt + c$, where h is height (in meters), t is time (in seconds), v is the initial upward velocity (in meters per second), and c is the initial height from the ground.

EXAMPLE 2   While standing on a platform 10 meters above the ground, Mike throws a ball upward with an initial velocity of 5 meters per second. About how long will the ball be in the air?

STEP 1   Use the formula. Let $c = 10$ and $v = 5$. You need to find how many seconds it will take for the ball to hit the ground. When the ball hits the ground, it will have a height of 0. Let $h = 0$.

STEP 2   Replace the variables with the given values. You can switch the sides of the equation so that 0 is on the right side of the equal sign.

$$h = -5t^2 + vt + c$$
$$-5t^2 + 5t + 10 = 0$$

STEP 3 Solve for t. First, divide each term by 5. Multiply each term by -1 to make it easier to factor.

$$-5t^2 + 5t + 10 = 0$$
$$-t^2 + t + 2 = 0$$
$$t^2 - t - 2 = 0$$

Factor the trinomial.

$$(t - 2)(t + 1) = 0$$

Set both factors equal to 0 and solve.

$$t - 2 = 0 \qquad t + 1 = 0$$
$$t = 2 \qquad\quad t = -1$$

STEP 4 Check the answers against the conditions in the problem. The problem asks for the time the ball is in the air. Time cannot be negative, so the negative solution, -1, is extraneous. Use the positive solution. The ball will be in the air approximately **2 seconds.**

Any application that involves multiplication has the potential to use quadratic equations. In the following example, you will need the formula $d = rt$, where $d =$ distance, $r =$ rate of speed, and $t =$ time.

EXAMPLE 3 A small plane flies 525 miles with a tailwind of 30 miles per hour. The plane then returns along the same path, flying against the wind. If the entire flight lasts 6 hours, how fast can the plane fly in still air?

STEP 1 Set up a chart to organize the information.

time = distance ÷ rate

	distance	rate	time
Upstream	525	$r + 30$	$\dfrac{525}{r + 30}$
Downstream	525	$r - 30$	$\dfrac{525}{r - 30}$

STEP 2 The total time is 6 hours. Write an equation.

$$\frac{525}{r + 30} + \frac{525}{r - 30} = 6$$

STEP 3 Solve. Multiply both sides by $(r + 30)(r - 30)$ to eliminate the fractions.

$$\frac{(r + 30)(r - 30)}{1} \cdot \left(\frac{525}{r + 30} + \frac{525}{r - 30} \right) = 6(r + 30)(r - 30)$$
$$525(r - 30) + 525(r + 30) = 6(r^2 - 900)$$
$$525r - 15{,}750 + 525r + 15{,}750 = 6(r^2 - 900)$$

Divide both sides by 6.

$$1{,}050r = 6(r^2 - 900)$$
$$175r = r^2 - 900$$

Move all the terms to one side of the equation.

The result is a quadratic equation.

$$r^2 - 175r - 900 = 0$$

Factor the equation.

$$(r - 180)(r + 5) = 0$$

Set each factor equal to 0, and find the two roots.

$$r - 180 = 0 \qquad r + 5 = 0$$
$$r = 180 \qquad\quad r = -5$$

Remember, you also can use the quadratic formula or completing the square to find the factors.

STEP 4 Decide which answer makes sense. It does not make sense for a plane to have a negative rate of speed. The positive root must be the answer.

The plane flies at a rate of **180 miles per hour** in still air.

Solve using any method.

1. If the sides of a square are increased by 4 inches, the area becomes 169 square inches. What is the area of the original square?

   (**Hint:** $A = s^2$, where s is the side of a square.)

2. The sum of a positive number and the square of that number is 240. What is the number?

3. The base of a triangle is 5 times its height. If the area of the triangle is 360 square centimeters, what is the base of the triangle?

   (**Hint:** Use the formula $A = \frac{1}{2}bh$.)

4. A rock is thrown upward from the roof of a building that is 20 meters from the ground. If the initial velocity is 15 meters per second, in about how many seconds will the rock hit the ground?

   (**Hint:** Use the formula $h = -5t^2 + vt + c$.)

5. A boat makes a 72-mile trip downstream and returns upstream to its starting point. If the rate of the current is 6 miles per hour and the entire trip took 9 hours, what is the boat's rate of speed in still water?

6. Lyn and Pat can row a boat 16 miles downstream and make the return trip upstream in a total time of 6 hours. If the current is 2 miles per hour, how fast can Lyn and Pat row the boat in still water?

7. When height is measured in feet and velocity is measured in feet per second, the following formula is used to show the height of a projectile over time:

   $$h = -16t^2 + vt + c$$

   A man stands on the roof of a tower that is 72 feet from the ground. He throws a tennis ball upward at a rate of 24 feet per second. In about how many seconds will the ball hit the ground?

8. A landscaper drew plans for a flower garden that is 12 feet by 20 feet. The owner of the lot wants to increase the length and the width of the garden by the same number of feet, increasing the total area of the garden by 320 square feet. What will be the dimensions of the new garden?

Quadratic Equations Review

Solve the problems below. When you finish, check your answers at the back of the book and correct any errors.

Write the quadratic equation in standard form, and find the values of a, b, and c.

1. $-9 = -4x^2 - 16x$

   $a =$
   $b =$
   $c =$

   $5n^2 = n + 21$

   $a =$
   $b =$
   $c =$

   $13 = -6a^2$

   $a =$
   $b =$
   $c =$

Write the solution set for each quadratic equation.

2. $2x(7x - 8) = 0$

   $(a - 9)(a - 4) = 0$

   $(b - 5)(b + 7) = 0$

3. $(n - 3)^2 = 0$

   $4(c - 1)(10c + 7) = 0$

   $(2m - 7)(3m + 2) = 0$

Factor. Then write the solution set for the equation.

4. $x^2 - 8x - 20 = 0$

   $3a^2 + 3a - 6 = 0$

   $m^2 - 15m + 54 = 0$

5. $b^2 + 11b + 18 = 0$

   $14x^2 + 6x = 0$

   $7n^2 + 68n - 20 = 0$

6. $5a^2 + a - 4 = 0$

   $2m^2 + m - 36 = 0$

   $6x^2 - 26x + 24 = 0$

7. $8x^2 + 26x - 7 = 0$

   $9m^2 - 3m - 2 = 0$

   $10a^2 - 79a + 63 = 0$

8. $45b^2 + 70b + 25 = 0$

   $8n^2 - 73n + 72 = 0$

   $24x^2 + 68x - 56 = 0$

Factor each expression completely.

9. $n^2 - 64$ $\qquad$ $25x^2 - 4$ $\qquad$ $49c^2 - 100$

10. $16a^2 - 8a + 1$ $\qquad$ $x^2 - 22x + 121$ $\qquad$ $81a^4 - 16$

11. $n^2 + 14n + 49$ $\qquad$ $25b^2 - 90b + 81$ $\qquad$ $128m^3 + 192m^2 + 72m$

Factor by grouping.

12. $10x^3 + 16x^2 - 15x - 24$ $\qquad$ $4m^3 - 14m^2 + 2m - 7$ $\qquad$ $5a^3 - 10a^2 + 20a - 40$

13. $2b^3 - 8b^2 + 9b - 36$ $\qquad$ $12x^3 - 4x^2 + 15x - 5$ $\qquad$ $5n^3 - 4n^2 - 20n + 16$

14. $84n^3 + 189n^2 + 12n + 27$ $\qquad$ $45a^3 + 5a^2 - 63a - 7$ $\qquad$ $24x^4 + 9x^3 + 168x^2 + 63x$

Simplify. Write excluded values in brackets.

15. $\dfrac{x - 2}{x^2 + 2x - 8}$ $\qquad$ $\dfrac{n^2 - 8n + 12}{n - 2}$ $\qquad$ $\dfrac{20a}{8a^2 + 32a}$

16. $\dfrac{b^2 + 7b - 8}{9b + 72}$ $\qquad$ $\dfrac{4n - 4}{n^2 + 8n - 9}$ $\qquad$ $\dfrac{x^2 - 1}{x^2 - 4x - 5}$

17. $\dfrac{x^3 - 8x^2 + 12x}{3x^3 - 24x^2 + 36x}$ $\qquad$ $\dfrac{n^2 + 4n - 21}{8n^2 - 8n - 48}$ $\qquad$ $\dfrac{2a^3 - 26a^2 + 80a}{4a^2 - 36a + 32}$

Add or subtract, and simplify.

18. $-2\sqrt{3} - 3\sqrt{3}$ $\qquad$ $-\sqrt{6} + 2\sqrt{54}$ $\qquad$ $2\sqrt{6} - \sqrt{24}$

19. $3\sqrt{27} - 3\sqrt{12}$ $\qquad$ $-2\sqrt{2} + 2\sqrt{8}$ $\qquad$ $-\sqrt{18} - \sqrt{18}$

20. $2\sqrt{2} - 3\sqrt{50}$ $\qquad$ $4\sqrt{63} - 5\sqrt{175}$ $\qquad$ $-2\sqrt{200} + 2\sqrt{128}$

Simplify.

21. $\dfrac{4\sqrt{8m^3}}{\sqrt{20m^2}}$ $\qquad$ $\dfrac{3\sqrt{2n}}{\sqrt{5n^3}}$ $\qquad$ $\dfrac{2x^2}{\sqrt{3x}}$

22. $\dfrac{3a + 3\sqrt{a^2}}{2\sqrt{7a}}$ $\qquad$ $\dfrac{4 + 3\sqrt{2x}}{\sqrt{10x}}$ $\qquad$ $\dfrac{5b - 2\sqrt{5b^2}}{4\sqrt{8b^4}}$

23. $\dfrac{4}{-4 - \sqrt{5}}$ $\qquad$ $\dfrac{5}{3 + \sqrt{3a}}$ $\qquad$ $\dfrac{4m}{\sqrt{3m^4} - \sqrt{m^4}}$

Solve by any method. Write *no solution* if the equation has no real number solution.

24. $(n + 4)^2 = 36$ $\qquad$ $(x + 2)^2 = 100$ $\qquad$ $(a + 3)^2 = -25$

25. $x^2 - 10x + 25 = 4$ $\qquad$ $n^2 + 20n + 100 = -9$ $\qquad$ $a^2 - 2a + 1 = 1$

26. $9m^2 - 30m + 25 = -16$ $\qquad$ $4a^2 + 12a + 9 = 4$ $\qquad$ $4x^2 + 28x + 49 = 1$

Solve by completing the square. Write irrational roots in simplest radical form.

27. $m^2 - 14m - 31 = 0$ $4a^2 - 8a - 95 = 0$ $8x^2 + 16x - 90 = 0$

28. $4n^2 + 3n - 9 = 0$ $8b^2 - 9b - 45 = 0$ $10a^2 + 17a + 6 = 0$

29. $3x^2 + 18x = 1$ $7n^2 - 14n - 47 = 2$ $2c^2 - 16c - 58 = -10$

30. $6b^2 + b - 5 = 10$ $2x^2 + 3x + 6 = 9$ $m^2 - 17m + 47 = -9$

Solve each equation for x using the quadratic formula. Write irrational roots in simplest radical form.

31. $3x^2 + 4x - 7 = 0$ $2x^2 - 6x - 18 = 0$ $5x^2 - 7x + 1 = 0$

32. $4x^2 + 2x - 1 = 0$ $6x^2 - 3 = 0$ $3x^2 - 8x + 4 = 0$

33. $5x^2 - 16 = -8x$ $x^2 - 6 = 4x$ $3x^2 = -1 - 8x$

34. $x^2 + 7x = -11$ $8x^2 = 5 + 4x$ $x^2 + 2 = -6x$

**Use the discriminant to determine the number of roots each equation has.
Write *one*, *two*, or *no solution*. Then use any method to find the roots.**

35. $x^2 = -12 - 7x$ $n^2 - 4n = -15$ $a^2 - 7 = -7$

36. $n^2 + 62 = 14n$ $4a^2 + 1 = 4a$ $c^2 + 8c = 32$

37. $7m^2 - 3m = 0$ $x^2 + 6x = -9$ $n^2 + 8n + 40 = 0$

Solve using any method.

38. A rectangular parking lot has an area of 1,400 square meters. If the length of the lot is 5 meters longer than its width, what are the dimensions of the lot?

39. The sum of a positive number and the square of that number is 90. What is the number?

40. The height of a falling object can be determined by the formula $h = -16t^2 + c$, where $h =$ height in feet, $t =$ time in seconds, and $c =$ initial height. A coin is dropped from a hot air balloon 256 feet in the air. About how many seconds will it take for the coin to hit the ground?

41. The height of a triangle is 4 inches shorter than the length of its base. If the area of the triangle is 48 square inches, what is the height of the triangle?

(**Hint:** Use the formula $A = \frac{1}{2}bh$.)

42. John can ride a bicycle 2 miles per hour faster than Matt. If both take the same 20-mile trip, John will arrive at the destination 30 minutes before Matt. What is John's average speed in miles per hour?

(**Hint:** 30 minutes $= 0.5$ or $\frac{1}{2}$ hour.)

43. When the sides of a square are increased by 8 centimeters, the area of the square increases by 144 square centimeters. What is the area of the original square?

GRAPHING QUADRATICS

When you graph any equation on the coordinate plane, the solution set of the graph forms a pattern of some kind. The solution set of a linear equation forms a line. The solution set of a quadratic equation forms a parabola. In this section you will see how the parts of an equation determine the exact shape of its graph. You will also see how to make changes to an equation to transform its graph.

Graphing Parabolas

In this lesson, you will learn to

- Graph quadratic equations.
- Find the vertex and the line of symmetry from the equation.

The graph of an equation in the form $y = mx + b$ is a line. Why does the graph form a straight line? A line is formed because for every x, there is one and only one y, and the change in y (the **rise**) and the change in x (the **run**) form a consistent relationship known as **slope.** Because slope is consistent, the values rise or fall in a linear pattern. For a review of graphing linear equations, see pages 65–66.

When an equation contains the square of x, the graph of the equation forms one of the shapes in a family of curves called **conic sections.** In this book, you will explore two of these shapes: parabolas and circles.

The graph of an equation in the form $y = ax^2 + bx + c$, where $a \neq 0$, is a **parabola.** A parabola has either a minimum, the lowest point on the curve, or a maximum, the highest point on the curve. This lowest or highest point is called a **vertex.** You can determine the vertex of a parabola from its equation.

Use the following formula to find the x-coordinate of the vertex of a parabola:

> ### x-coordinate of the Vertex of a Parabola
> If $y = ax^2 + bx + c$ and $a \neq 0$, then the x-coordinate of the vertex is $x = -\frac{b}{2a}$.

Once you know the x-coordinate of the vertex, you can substitute that value for x in the original equation and solve for the y-coordinate of the vertex.

EXAMPLE 1 What is the vertex of $y = -x^2 + 2x$?

STEP 1 Find the x-coordinate of the vertex. Use the formula $x = -\frac{b}{2a}$.

$$x = -\frac{b}{2a} = -\frac{2}{2(-1)}$$
$$= \frac{-2}{-2}$$
$$= 1$$

STEP 2 Find the y-coordinate of the vertex. Substitute 1 for x in the original equation.

$$y = -x^2 + 2x = -1^2 + 2(1)$$
$$= -1 + 2$$
$$= 1$$

The vertex of the graph is located at **(1, 1).**

In a quadratic equation of the form $y = ax^2 + bx + c$, where $a \neq 0$, the coefficient a can be either positive or negative. You can use this fact to help you graph.

- If a is positive, $(a > 0)$, then the parabola opens upward.
- If a is negative, $(a < 0)$, then the parabola opens downward.

Another feature of a parabola that is helpful for graphing is that a parabola is symmetrical. A shape is symmetrical when one side is the mirror image, or reflection, of the other side. If you fold a shape in half and the two halves line up perfectly, then the shape is symmetrical. The line that you folded is called the **line of symmetry.**

A parabola's line of symmetry passes through its vertex. Points that share the same y-coordinate are symmetric points. Symmetric points lie the same distance from the line of symmetry. On the graph shown, the points $(1, -6)$ and $(5, -6)$ are symmetric. The points $(2, -3)$ and $(4, -3)$ are also symmetric. A parabola contains infinitely many symmetric points.

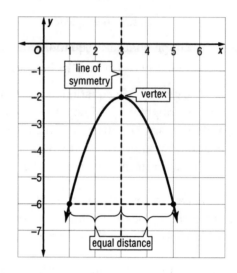

When you are given the equation of a parabola, you can use what you know about the vertex and the line of symmetry to sketch the parabola's graph.

EXAMPLE 2   Graph the equation: $y = 2x^2 + 8x + 9$

You know the parabola will open upward because a is positive ($a = 2$).

STEP 1 Find the x-coordinate of the vertex.

Use the formula $x = -\dfrac{b}{2a}$ to find x.

Use the values $a = 2$ and $b = 8$.

$$x = -\frac{b}{2a} = -\frac{8}{2(2)} = -\frac{8}{4} = -2$$

STEP 2 Find the y-coordinate of the vertex.

Substitute –2 for x in the original equation, and solve for y.

The vertex is located at **(–2, 1).**

$$y = 2x^2 + 8x + 9 = 2(-2)^2 + 8(-2) + 9$$
$$= 2(4) - 16 + 9$$
$$= 8 - 16 + 9$$
$$= 1$$

STEP 3 Plot the vertex on the graph, and draw a vertical line of symmetry through the vertex.

STEP 4 Find two other points on the graph. Look at the equation $y = 2x^2 + 8x + 9$. If $x = 0$, then $y = 9$. One point on the graph is $(0, 9)$. Plot the point.

Choose a value for x that is on the same side of the vertex as the point $(0, 9)$.

If $x = -1$, then $y = 3$.
A second point on the graph is $(-1, 3)$.
Plot that point.

$$y = 2x^2 + 8x + 9$$
$$y = 2(-1)^2 + 8(-1) + 9$$
$$y = 2 - 8 + 9$$
$$y = 3$$

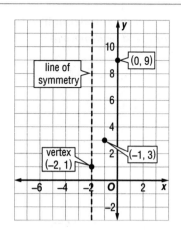

STEP 5 Plot symmetric points, and graph the equation.

Symmetric points have the same y-coordinates, and they are the same distance from the line of symmetry.

The point symmetric to $(0, 9)$ is $(-4, 9)$.

The point symmetric to $(-1, 3)$ is $(-3, 3)$.

Plot the points $(-4, 9)$ and $(-3, 3)$. Then carefully sketch the graph through the points.

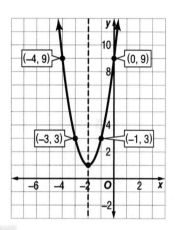

Note: The plotted points are only a sampling of the solution set for the equation. Remember, every point on the parabola is part of the solution set.

Recall from the previous chapter that the roots of a quadratic equation are the points where the parabola crosses the x-axis. One way to solve a quadratic equation is to set y equal to 0 and sketch its graph. The x-intercepts are the solutions.

 Think About It: The parabola in Example 2 does not cross the x-axis. What does this tell you about its roots? This tells you that the quadratic equation has no real roots. What is the graph of a quadratic equation with exactly 1 root? This graph is a parabola whose vertex lies on the x-axis.

Graph of a quadratic equation with exactly **1 root**

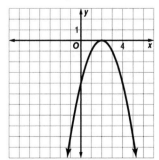

Graph of a quadratic equation with **2 roots**

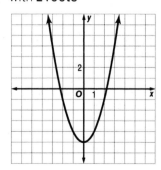

For each equation, find the vertex of the graph. Then write whether the graph opens *upward* or *downward*.

1. $y = -x^2 + 4x - 5$ $\qquad\qquad$ $y = 2x^2 - 4x + 5$ $\qquad\qquad$ $y = 7x^2 + 42x + 12$

2. $y = 3x^2 - 12x + 1$ $\qquad\qquad$ $y = -5x^2 + 20x - 6$ $\qquad\qquad$ $y = -4x^2 - 8x + 3$

3. $y = -2x^2 + 4x - 3$ $\qquad\qquad$ $y = 6x^2 + 12x - 1$ $\qquad\qquad$ $y = 3x^2 - 24x + 20$

For each equation, find the coordinates of the vertex. Plot the vertex, and draw the line of symmetry. Then plot symmetric points and sketch the graph.

4. $y = x^2 - 4x + 3$ $\qquad\qquad\qquad\qquad$ $y = -x^2 - 4x - 8$

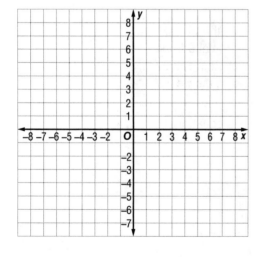 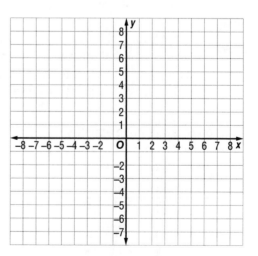

5. $y = x^2 + 4x + 5$

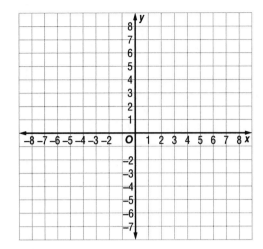

$y = x^2 + 2x + 4$

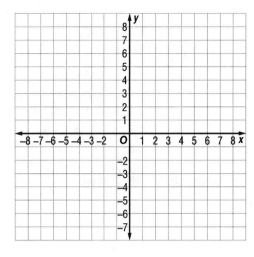

6. $y = -2x^2 + 4x - 5$

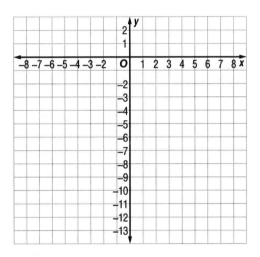

$y = -3x^2 + 6x - 4$

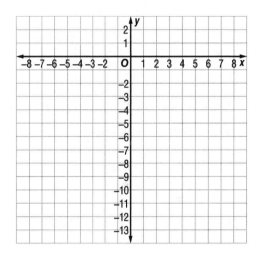

7. $y = x^2 + 4x + 8$

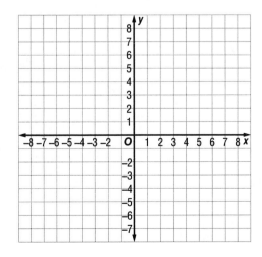

$y = 2x^2 - 8x + 7$

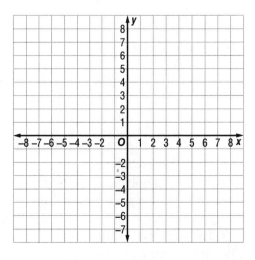

Patterns in Graphing Equations

In this lesson, you will learn to

- Recognize the patterns formed by the graphs of equations.

You have noticed that the graphs of some equations create particular shapes. Mastering these equations and their forms will help you with higher mathematics and functions. As you study the following graphs, think about why the equation forms that pattern. Pay close attention to whether a positive or negative sign appears in front of the x-variable.

Linear Equations

$y = x$

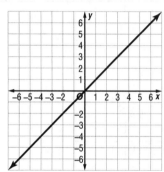

$y = -x$

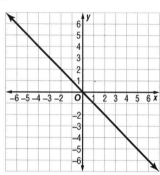

What do you notice about these linear graphs?

- Both are straight lines.
- The first graph has a positive slope; it slants upward from left to right.
- The second equation has a negative sign before the x-variable. Its graph has a negative slope; it slants downward from left to right.

Quadratic Equations

$y = x^2$

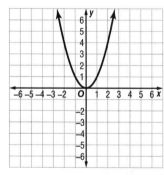

$y = -x^2$

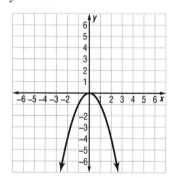

What do you notice about these quadratic graphs?

- Both are parabolas.
- The first graph has a minimum point and opens upward.
- The second equation has a negative sign before the x-variable. Its graph has a maximum point and opens downward.
- For every y-coordinate value, there are two x-coordinate values, except at the vertex.

All graphs of x raised to an even power form a parabola.

The graph of the equation $y = x^4$ is at the right. Why is it a quadratic equation? Because $x^4 = (x^2)^2$.

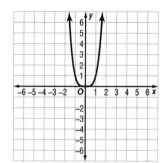

Absolute-Value Equations

$y = |x|$

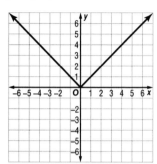

$y = -|x|$

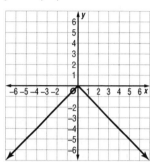

What do you notice about these absolute-value graphs?

- Both form a V shape.
- The first graph has a minimum point and opens upward.
- The second equation has a negative sign before the absolute value of the x-variable. Its graph has a maximum point and opens downward.
- For every y-coordinate value, there are two x-coordinate values, except at the vertex.

Square-Root Equations

$y = \sqrt{x}$

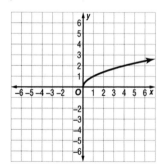

$y = -\sqrt{x}$

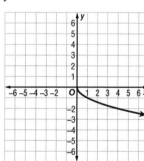

What do you notice about these square-root graphs?

- A square-root graph looks like a parabola turned on its side and cut in half along its line of symmetry. If you combined both graphs into one, it would form a parabola that opens to the right.

Note: If there were negative signs inside the radicals ($y = \sqrt{-x}$ and $y = -\sqrt{-x}$), the two pieces would form a parabola that opens to the left.

Cubic Equations

$y = x^3$

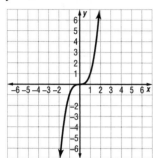

$y = -x^3$

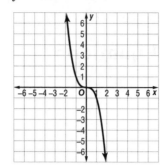

What do you notice about these cubic graphs?

- A graph of a cubic equation looks like a parabola with a twist. One half of the parabola is above the x-axis and the other half is below. If there is a negative sign before the x-variable, the curve moves downward from left to right.

Note: All graphs of x raised to an odd power greater than 1 form this same shape.

Rational Equations

$y = \dfrac{1}{x}$

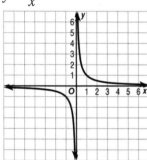

$y = -\dfrac{1}{x}$

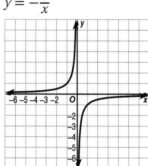

What do you notice about these rational-equation graphs?

- Because division by 0 is undefined, x cannot equal 0. Because the numerator is not equal to 0, y cannot equal 0. As a result, the graph moves close to the values that would make x or y equal 0, but the graph never touches those values.
- If there is a negative sign before the x-variable, the y-values will have the opposite sign of the x-values. Therefore, the graph will be above the x-axis where x is negative and below the x-axis where x is positive.

This short lesson contains the basic forms of graphs that you will see again and again. Of course, equations are often more complicated than those shown here, but these basic shapes remain the same.

To determine the basic shape of a graph, solve the equation for y and write it in standard form. Then look at the first term. Ask yourself: "What is happening to the x-variable? Is it squared or cubed? Is it part of a fraction or a radical? Is it positive or negative?" Then think about the characteristics of its graph.

Understanding graphs will help you in your work with transformations and functions.

Each graph was created using a graphing calculator. Use the basic forms you have learned to match each equation to its graph.

1. $y = x - 5$

A.

F.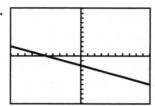

2. $y = |x + 2|$

3. $y = -x^2 + 3$

B.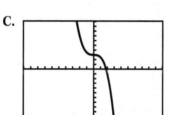

G.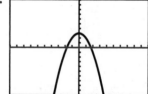

4. $y = \sqrt{x + 4}$

5. $y = -\frac{2}{5}x - 2$

C.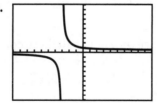

H.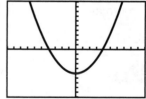

6. $y = \frac{2}{x + 3}$

7. $y = -2|x - 4|$

D.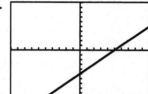

I.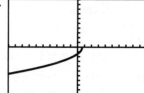

8. $y = \frac{1}{3}x^2 - 5$

9. $y = -\frac{1}{2}x^3 + 3$

E.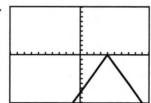

J.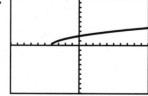

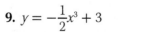

10. $y = -\sqrt{2 - 3x}$

Transformations

In this lesson, you will learn to

- Apply translations, reflections, and rotations to figures on a coordinate plane.
- Understand how changes to an equation will affect its graph.

In the language of transformations, a rule is applied to an original shape, called the **preimage,** to create a final shape, called the **image.** The preimage could be a geometric shape or the graph of an equation or function.

You can transform a basic geometric shape by applying changes to the x- and y-coordinates of its vertices. Example 1 shows the translation of a figure. A **translation** slides a graph up, down, right, or left. You can translate a figure by graphing the preimage and then counting spaces to move the vertices. Label the new vertices using the prime symbol (′).

EXAMPLE 1 Four-sided figure *QRST* has vertices at $Q(-1, 5)$, $R(1, 3)$, $S(-1, -1)$ and $T(-4, 3)$. Translate the figure right 3 units and down 2 units.

Graph *QRST*.

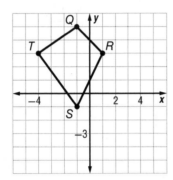

Shift each vertex right 3 units and down 2 units.

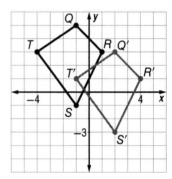

You can also find the new vertices by adding 3 to each x-coordinate and subtracting 2 from each y-coordinate. Adding shifts up or right, and subtracting shifts left or down. For example, if you apply the translation $(+3, -2)$ to $Q(-1, 5)$, the new point is $Q'(2, 3)$.

A **reflection** flips a figure across an axis, line, or point. In other words, it creates a mirror image of the preimage in a new location.

To reflect a figure across the y-axis, change the sign of all the x-coordinates.

EXAMPLE 2 Figure *ABCD* has vertices at $A(5, 1)$, $B(2, -1)$, $C(2, -2)$ and $D(5, -4)$. Reflect the figure across the y-axis.

Graph *ABCD*.

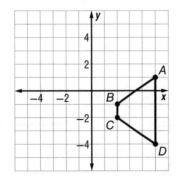

Change the sign of each x-coordinate, and then graph:

$A'(-5, 1)$

$B'(-2, -1)$

$C'(-2, -2)$

$D'(-5, -4)$

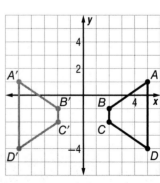

To reflect across the *x*-axis, change the sign of all the *y*-coordinates.

EXAMPLE 3 Figure *JKLM* has vertices at *J*(5, −2), *K*(1, 0), *L*(0, −3) and *M*(4, −5). Reflect the figure across the *x*-axis.

Graph *JKLM*.

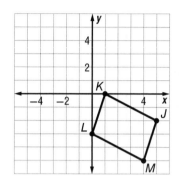

Change the sign of each *y*-coordinate, and then graph:

J′(5, 2)

K′(1, 0)

L′(0, 3)

M′(4, 5)

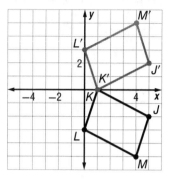

The final type of transformation discussed in this book is **rotation.** A rotation turns the graph about a point. In this book, a rotation will always be about the origin (0, 0). Imagine sticking a pin through the origin and then actually turning the coordinate plane counterclockwise 90° (a quarter turn), 180° (a half turn), or 270° (a three-quarter turn).

This is actually one method for solving a rotation problem. You can physically turn the coordinate plane and then draw the graph. You can also graph a rotation by making changes to the *x*- and *y*-coordinates.

Mathematicians have agreed that any rotation stated as a positive number turns the graph counterclockwise. A clockwise rotation is measured in negative degrees.

For each ordered pair (*x*, *y*), represented by (A, B),

- The rule for a 90° counterclockwise rotation about the origin is (A, B) → (−B, A).
- The rule for a 180° counterclockwise rotation about the origin is (A, B) → (−A, −B).
- The rule for a 270° counterclockwise rotation about the origin is (A, B) → (B, −A).

EXAMPLE 4 Triangle *DFG* has vertices at *D*(−5, −3), *F*(−2, −2), *G*(0, −5). Rotate triangle *DFG* 270° about the origin.

Graph *DFG*.

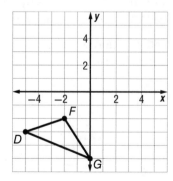

The rule for 270° is (A, B) ⟶ (B, −A).

Switch the order of the coordinates, and change the sign of A.

D(−5, −3) ⟶ *D*′(−3, 5)

F(−2, −2) ⟶ *F*′(−2, 2)

G(0, −5) ⟶ *G*′(−5, 0)

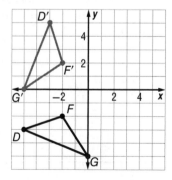

Think About It: A clockwise rotation is expressed as a negative number of degrees. How could you find a clockwise rotation? Think of it as a counterclockwise rotation. Rotating a graph 90° clockwise is the same as rotating it 270° counterclockwise. Therefore, you can use the rule for a 270° counterclockwise rotation.

Perform the transformations as directed.

1. Graph figure *GRLV* at *G*(−5, 4), *R*(−4, 4), *L*(−2, 5), and *V*(−3, 0). Then translate the figure right 7 units and down 2 units.

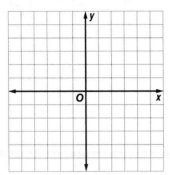

Graph figure *LNQ* at *L*(3, 3), *N*(5, −2), and *Q*(1, 0). Then translate the figure left 4 units and up 1 unit.

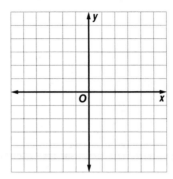

2. Graph figure *CKLS* at *C*(2, −1), *K*(5, 3), *L*(1, 5), and *S*(−1, 3). Then reflect *CKLS* across the *x*-axis.

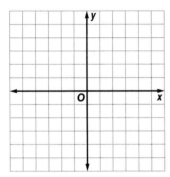

Graph figure *KHZU* at *K*(3, 1), *H*(−2, 2), *Z*(−1, −2), and *U*(3, −4). Then reflect *KHZU* across the *y*-axis.

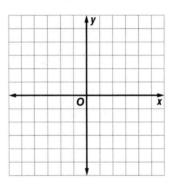

3. Graph figure *BYNV* at *B*(1, 0), *Y*(1, 5), *N*(3, 5), and *V*(4, 0). Then rotate the figure 90° about the origin.

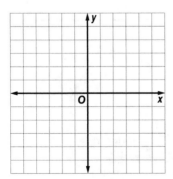

Graph figure *MQK* at *M*(−2, −3), *Q*(−3, 1), *K*(−1, 0). Rotate the figure −180° about the origin.

(**Hint:** A 180° rotation is the same clockwise and counterclockwise.)

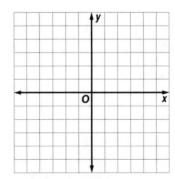

Think of the graph of an equation as a transformation of the parent equation. The parent equation is the basic form for the type of equation. For example, $y = x^2$ is the basic form for a quadratic equation. You learned basic forms in the last lesson.

You can take any graph on the coordinate plane and transform it in any combination of the following ways:

- A **translation** slides the graph up, down, right, or left.
- A **reflection** flips the graph over an axis, creating a mirror image.
- A **non-rigid transformation** stretches or shrinks the graph by changing the steepness of the curves.

You can create a transformation by making changes to parts of an equation.

Change the sign here to reflect the graph across the x-axis.

Add a number here to shift the graph to the left. Subtract a number to shift the graph to the right.

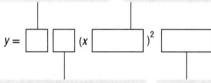

Insert a number here to change the steepness of the curves. A fraction less than 1 widens the graph. A number greater than 1 makes the graph narrower.

Add a number here to shift the graph up. Subtract a number to shift the graph down.

Notice how the changes to the parent equation $y = x^2$ change the graphs below.

STEP 1 Graph the parent function.

$y = x^2$

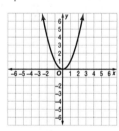

STEP 2 Reflect the graph across the x-axis.

$y = -x^2$

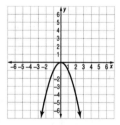

STEP 3 Translate the graph up 3 units.

$y = -x^2 + 3$

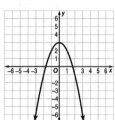

STEP 4 Make the graph less steep, or wider.

$y = -\frac{1}{2}x^2 + 3$

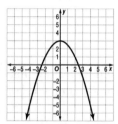

STEP 5 Translate the graph 2 units right.

$y = -\frac{1}{2}(x - 2)^2 + 3$

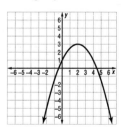

Do you see how changes to these four parts of the equation change the graph of the parent equation?

You can perform these same changes with any parent equation.

Choose the equation that represents each transformation from the parent graph.

For problems 4–8, refer to this graph of the parent equation $y = x^2$.

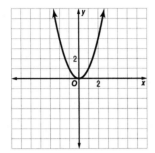

4. a. $y = x^2 - 5$
   b. $y = x^2 + 5$

5. a. $y = x^2 + 4$
   b. $y = (x + 4)^2$

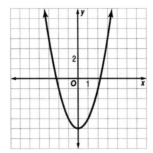

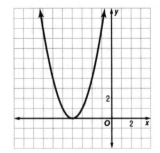

6. a. $y = -\frac{1}{3}x^2$
   b. $y = -3x^2$

7. a. $y = -(x - 2)^2$
   b. $y = -2x^2$

8. a. $y = \frac{1}{4}(x + 3)^2$
   b. $y = 2x^2 - 5$

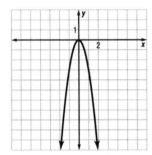

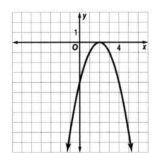

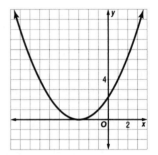

For problems 9 and 10, refer to this graph of the parent equation $y = x^3$.

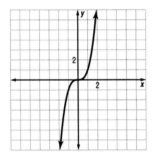

9. a. $y = \frac{1}{6}x^3$
   b. $y = x^3 - 6$

10. a. $y = x^3 - 4$
   b. $y = (x + 4)^3$

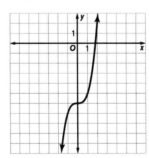

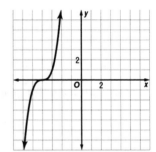

For problems 11 and 12, refer to this graph of the parent equation $y = |x|$.

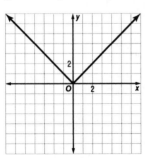

11. a. $y = -|x| - 5$
   b. $y = -|x - 5|$

12. a. $y = 2|x|$
   b. $y = \frac{1}{2}|x|$

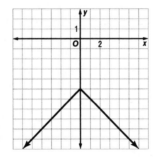

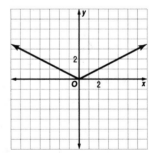

Circles

In this lesson, you will learn to

- Write the equation of a circle and graph the circle.
- Solve for the radius or center of a circle when the equation is known.

On a coordinate plane, a **circle** consists of the set of all points that lie an equal distance, the **radius,** from a fixed point on the plane, the **center.** In other words, every point on the circle is the same distance from the center.

You know how to find the distance between two points (x_1, y_1) and (x_2, y_2) on a coordinate plane.

$$d = \sqrt{(x_2 - x_1)^2 + (y_2 - y_1)^2}$$

You can use this formula to create an equation for a circle. The distance from the center of the circle to any point on the curve must be the radius (r) of the circle. Mathematicians have chosen to represent the center of a circle by the ordered pair (h, k). Any point on the circle itself can be represented by (x, y).

Substitute these values into the distance formula.

$$r = \sqrt{(x - h)^2 + (y - k)^2}$$

You can eliminate the radical sign by squaring both sides of the equation. This is the standard form for the equation of a circle.

$$r^2 = (x - h)^2 + (y - k)^2$$

Standard Form for the Equation of a Circle	$(x - h)^2 + (y - k)^2 = r^2$

You can simplify this equation if the center is at the origin $(0, 0)$. Replace (h, k) with $(0, 0)$.

$$(x - 0)^2 + (y - 0)^2 = r^2$$
$$x^2 + y^2 = r^2$$

Sometimes the radius and the center of a circle are all you need to graph the circle. This is true when the radius and the center are easy numbers to work with on the coordinate plane.

EXAMPLE 1 Graph the equation of a circle with center at $(-3, 4)$ and a radius of 3.

STEP 1 Locate the center of the circle.

STEP 2 Determine four points on the circle by counting 3 units up, down, right, and left.

STEP 3 Sketch a smooth curve through all four points.

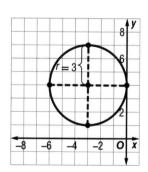

You can use the standard form equation to write the equation for a circle if you are given the center and radius of the circle.

EXAMPLE 2 Write the equation of the circle shown on the graph.

STEP 1 Examine the graph. The center of the circle is (1, −1). Count from the center to a point directly above it on the graph. The radius is 5.

STEP 2 Replace the variables in the standard form equation and simplify.

$$(x - h)^2 + (y - k)^2 = r^2$$
$$(x - 1)^2 + (y - -1)^2 = 5^2$$
$$(x - 1)^2 + (y + 1)^2 = 25$$

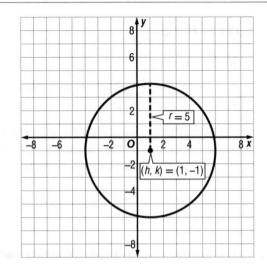

Using algebra, you can determine the center and the radius of a circle from its equation.

EXAMPLE 3 The equation of a circle is $(x - 8)^2 + (y + 3)^2 = 45$. Find the center and radius of the circle.

STEP 1 Find the center (h, k). Compare the equation of the circle with the standard form.

You can see that $h = 8$.

The variable k must equal −3. Do you see why?

If $y - k = y + 3$, then $-k = 3$ and $k = -3$.

$$(x - h)^2 + (y - k)^2 = r^2$$
$$(x - 8)^2 + (y + 3)^2 = 45$$

Center of the circle: $(h, k) = (8, -3)$

STEP 2 Find the radius.

The radius is a measure of distance, so you only need the positive root.

Simplify.

$$r^2 = 45$$
$$\sqrt{r^2} = \sqrt{45}$$
$$r = \sqrt{45}$$
$$r = 3\sqrt{5}$$

Radius of the circle: $r = 3\sqrt{5}$

For each equation of a circle, find the center and radius.

1. $(x + 14)^2 + (y - 10)^2 = 9$ $(x + 7)^2 + (y + 6)^2 = 36$ $(x - 7)^2 + (y + 11)^2 = 25$

2. $(x - 8)^2 + (y - 5)^2 = 8$ $(x + 6)^2 + (y + 11)^2 = 52$ $(x + 8)^2 + (y - 15)^2 = 12$

3. $\left(x - \frac{19}{2}\right)^2 + \left(y + \frac{9}{2}\right)^2 = 16$ $(x - 4\sqrt{7})^2 + \left(y - \frac{23}{2}\right)^2 = 36$ $(x - \sqrt{35})^2 + (y + \sqrt{13})^2 = 54$

For each graph, write the center, radius, and equation of the circle.

4.
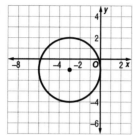
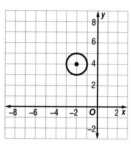

5.
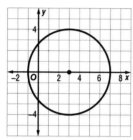

Synthesis   Because circles are geometric figures, you can determine their area and circumference. **Circumference** is the distance around a circle. To find circumference and area, you will use the constant **pi (π)**, which has a value of about 3.14. The two formulas are as follows:

Circumference of a circle	$C = 2\pi r$ or $C = \pi d$
Area of a circle	$A = \pi r^2$

Use the formulas for circumference and area to solve the problems.

6. Which of the following represents the area of a circle with the equation $(x + 7)^2 + (y + 2)^2 = 49$?

   a. 7π

   b. 14π

   c. 49π

7. Find the area and the circumference of the circle shown on the graph below. Use 3.14 for π.

   Area: _____

   Circumference: _____

8. The circumference of a circle is 6π. The center of the circle is (4, 12). Which of the following represents the equation of the circle?

   (**Hint:** Use the circumference formula to find the radius.)

   a. $(x - 4)^2 + (y - 12)^2 = 6$

   b. $(x - 4)^2 + (y - 12)^2 = 9$

   c. $(x - 4)^2 + (y - 12)^2 = 36$

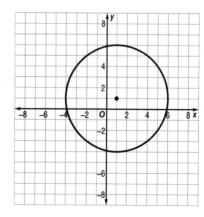

Graphing Quadratics Review

Solve the problems below. When you finish, check your answers at the back of the book and correct any errors.

For each equation, find the vertex of the graph. Then write whether the graph opens *upward* or *downward*.

1. $y = -\frac{1}{2}x^2 + 2x + 2$ $\qquad$ $y = x^2 + 4x + 1$ $\qquad$ $y = 2x^2 - 12x + 21$

2. $y = x^2 - 5$ $\qquad$ $y = -2x^2 + 16x - 29$ $\qquad$ $y = -x^2 - 8x - 14$

3. $y = 2x^2 - 4x + 1$ $\qquad$ $y = -2x^2 - 4x$ $\qquad$ $y = -\frac{1}{2}x^2 + 2x + 3$

For each equation, find the coordinates of the vertex. Then sketch the graph of the equation.

4. $y = x^2 + 4x + 5$ $\qquad\qquad\qquad\qquad$ $y = -x^2 - 2x - 4$

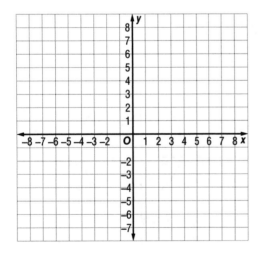 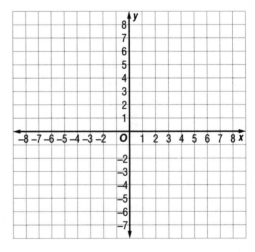

5. $y = x^2 + 2x - 1$

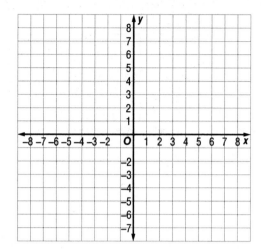

$y = -\dfrac{1}{4}x^2 - x - 2$

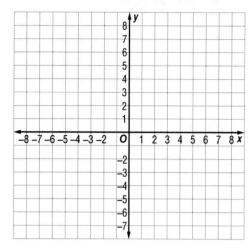

6. $y = x^2 - 4x + 4$

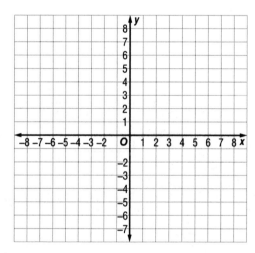

$y = x^2 - 6x + 8$

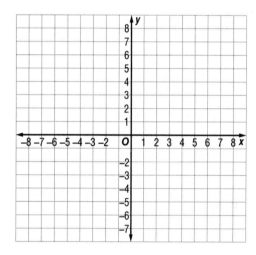

7. $y = -x^2 + 6x - 5$

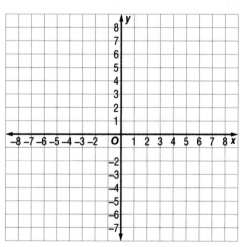

$y = -\dfrac{2}{3}x^2 - 4x - 8$

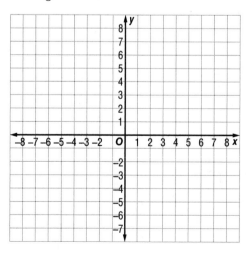

Each graph was created using a graphing calculator. Use the basic forms you have learned to match each equation to its graph.

8. $y = -x^2 + 5$

A.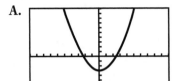

F.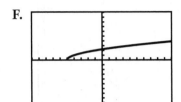

9. $y = -\frac{3}{4}|x - 2|$

10. $y = -\sqrt{4 - x}$

B.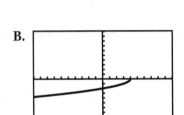

G.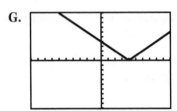

11. $y = -\frac{1}{3}x^3 - 5$

12. $y = x - 3$

C.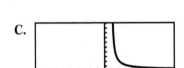

H.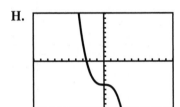

13. $y = \frac{2}{x - 1}$

14. $y = -\frac{1}{4}x - 5$

D.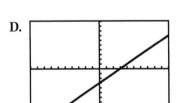

I.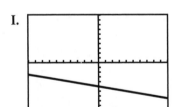

15. $y = \sqrt{x + 5}$

16. $y = \frac{1}{2}x^2 - 3$

E.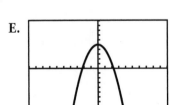

J.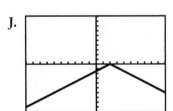

17. $y = |x - 4|$

18. Translate figure *DFPY* left 5 units and up 2 units.

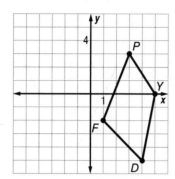

Translate figure *ZAL* right 4 units and down 1 unit.

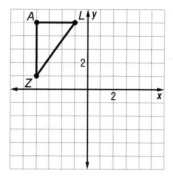

19. Reflect figure *JRKX* across the *x*-axis.

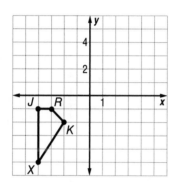

Reflect figure *WBXE* across the *y*-axis.

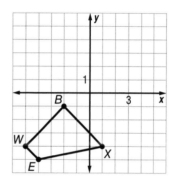

20. Rotate figure *ZTFR* 180° about the origin.

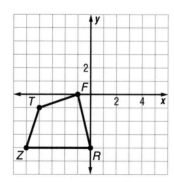

Rotate figure *SKN* −90° about the origin.

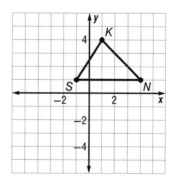

Choose the equation that represents each transformation from the parent graph.

For problems 21 and 22, refer to this graph of the parent equation $y = |x|$.

21. a. $y = |-x| + 4$
  b. $y = -|x - 4|$

22. a. $y = 2|x| - 3$
  b. $y = \frac{1}{2}|x| + 3$

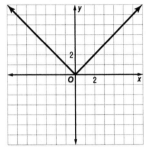

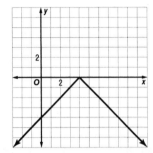

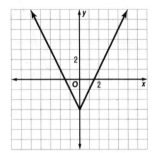

For problems 23 and 24, refer to this graph of the parent equation $y = x^3$.

23. a. $y = 3x^3$
  b. $y = \frac{1}{3}x^3$

24. a. $y = x^3 + 5$
  b. $y = (x + 5)^3$

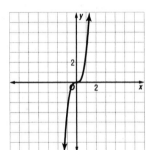

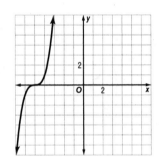

For each equation of a circle, find the center and radius.

25. $(x - 7)^2 + (y + 2)^2 = 36$
  $(x + 11)^2 + (y + 8)^2 = 49$
  $(x + 8)^2 + (y - 1)^2 = 1$

26. $(x - 1)^2 + (y + 7)^2 = 63$
  $(x - 9)^2 + (y + 16)^2 = 3$
  $(x + 4)^2 + (y - 8)^2 = 24$

27. $x^2 + (y - 1)^2 = 64$
  $(x - 7)^2 + (y - 3)^2 = 117$
  $(x + 3)^2 + y^2 = 12$

For each graph, write the center, radius, and equation of the circle.

28.

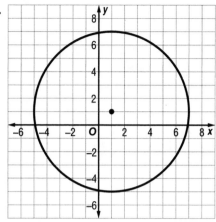

29.

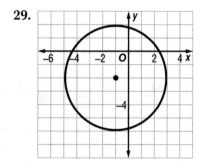

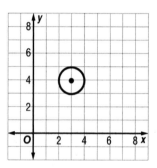

For each equation, graph the circle.

30. $(x + 1)^2 + (y - 2)^2 = 16$

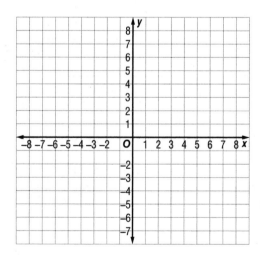

$(x - 1)^2 + y^2 = 25$

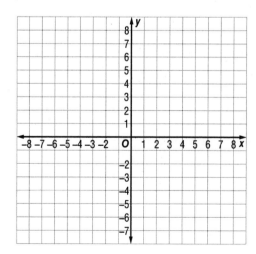

EXTENSION TOPICS

In any field of study, there is always more to learn. In this section, you will learn about three topics that go beyond the basics of algebra. As you read about functions, systems of inequalities, and trigonometric ratios, think about how these concepts connect with the ideas you have already mastered.

Function Basics

In this lesson, you will learn to

- Find function values or outputs.
- Find the domain and range of a function.
- Determine whether a graph represents a function.
- Combine functions.

A **function** describes the relationship between two sets: the domain and the range. It can be helpful to think of a function as a machine. When a value from the domain enters the machine, a value from the range comes out.

Consider the table of inputs and outputs at the right. What process would cause each input (x) to yield its corresponding output $f(x)$?

The relation that connects these two sets of numbers is $f(x) = 3x - 1$. Read $f(x)$ as "f of x."

Inputs x	Outputs $f(x)$
1	2
2	5
3	8
4	11

You can think of a function as a machine.

For any value of x, the machine creates exactly one $f(x)$.

In the next example, you will see how functions are used in practice. Although any variable can be used to label a function, mathematicians generally reserve the variables f, g, and h for functions.

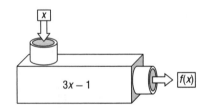

EXAMPLE 1   For $f(x) = x^3 + 2$, find $f(-2)$ and $f(3)$.

You need to find the value of the function when $x = -2$ and when $x = 3$. Substitute and solve.

$$f(x) = x^3 + 2$$
$$f(-2) = (-2)^3 + 2$$
$$f(-2) = -6$$

$$f(x) = x^3 + 2$$
$$f(3) = (3)^3 + 2$$
$$f(3) = 29$$

If -2 is substituted for x in the function $f(x) = x^3 + 2$, the output is **−6.** If the input is 3, the output is **29.**

Find the function values as indicated.

1. $f(x) = -x^2 - 4$ $f(0)$ $f(-4)$ $f(5)$

2. $g(x) = 3x^2 + 2x - 1$ $g(-3)$ $g(1)$ $g(3)$

3. $h(x) = -2|3x|$ $h(9)$ $h(-4)$ $h(-1)$

4. $f(x) = 3x - 5$ $f(2)$ $f(0)$ $f(-3)$

    The inputs and outputs of a function must be real numbers. If a particular value for the variable will result in an output that is not a real number, you must exclude that value from the set of inputs. Remember, the **domain** is the set of inputs and the **range** is the set of outputs.

    What situations must you watch out for? Pay close attention to fractions. Inputs that cause a denominator to equal 0 cannot be part of the domain of a function. Also, inputs that cause a negative radicand in an even root must be excluded from the domain.

EXAMPLE 2   Find the domain of $f(x) = \dfrac{2}{x - 5}$.

What value will cause the denominator to equal 0? $x - 5 = 0$
Because x must not equal 5, the domain is $\{x \mid x \neq 5\}$. $x = 5$

The statement $\{x \mid x \neq 5\}$ is read "the set of all x such that x does not equal 5."

    Set builder notation is written in brackets as shown above. You can also write the domain using interval notation: $(-\infty, 5) \cup (5, \infty)$. For a review of interval notation, see page 50. In this lesson, you will use set builder notation to write domain and range.

EXAMPLE 3   Find the domain of $g(x) = 3 + \sqrt{2 - x}$.

This function involves a radicand of an even root. The radicand $2 - x \geq 0$
must be greater than or equal to 0. $-x \geq -2$
Write an inequality and solve. $x \leq 2$

Remember to reverse the sign of an inequality if you multiply or divide both sides by a negative number.

The value of x must be less than or equal to 2. The domain is $\{x \mid x \leq 2\}$, which reads: "the set of all x such that x is less than or equal to 2."

EXAMPLE 4 Find the domain of $h(x) = x^2 + |x|$.

This expression doesn't have a denominator or a radicand. There are no exclusions to the domain. Any number can be substituted for x. Therefore, the domain is **the set of all real numbers,** which can be written as $\mathbb{R}$.

Write the domain of each function. Use set builder notation. Write $\mathbb{R}$ to indicate the set of all real numbers.

5. $f(x) = 5x + 2$ $g(x) = |x^2 - 5|$ $h(x) = 3 - \dfrac{1}{x + 4}$

6. $f(x) = \dfrac{x + 5}{x - 2}$ $g(x) = \dfrac{1}{x^2 - 25}$ $h(x) = \sqrt{6 - x}$

7. $f(x) = \dfrac{\sqrt{x - 4}}{x + 2}$ $g(x) = \dfrac{3}{x^2 + x - 6}$ $h(x) = \frac{1}{2}|x + 4|$

   The definition of a function states that for every input, there can be only one output. Suppose x represents each input and y represents each output. Then each pair of values (input, output) forms an ordered pair (x, y). Using these ordered pairs, you can graph any function just as you graphed equations.

   This raises a question: Is every graph a function? No. A graph is not a function if any x-coordinate can be paired with more than one y-coordinate. You can determine whether a graph represents a function by applying the **vertical line test.**

> **Vertical Line Test**   If it is possible for a vertical line to cross a graph at more than one point, then the graph does not represent a function.

EXAMPLE 5 Determine which of the graphs below represents a function.

Examine the graphs. For each graph, think: *Is it possible to draw a vertical line that passes through two or more points?*

No matter where you draw the vertical line in the first graph, it will pass through only one point.

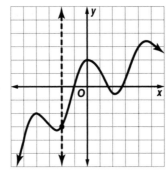

The vertical line crosses at only one point. **This is a function.**

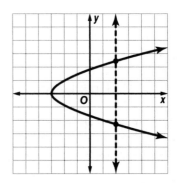

The vertical line crosses at two points. **This is not a function.**

Do not be fooled by a graph with an open circle. At first glance, it seems that the vertical line passes through two points on the graph at the right. Remember, however, that an open circle means that the point is not included in the graph. The solid circle, on the other hand, is a point on the graph.

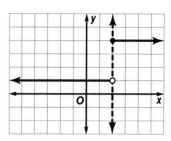

This is an example of a **discontinuous function.** The graph seems to jump to another spot on the grid.

This is a function.

Determine whether each graph represents a function. Write *Yes* or *No.*

8.

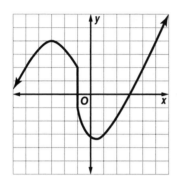 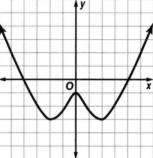 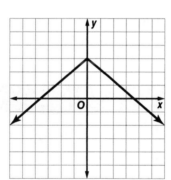

9.

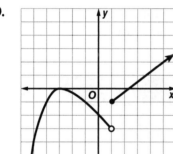 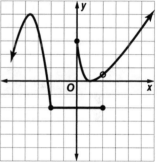

You can find the domain and range of a function from its graph. The domain consists of the set of all *x*-values. The range consists of the set of all *y*-values.

EXAMPLE 6   Determine the domain and the range of the function from the graph.

STEP 1   Find the domain. Look at the horizontal change in the graph. It stretches from −4 to 3 on the *x*-axis. The open circle means that 3 is not included.

STEP 2   Find the range. Look at the vertical change in the graph. The line stretches from −1 at the lowest point to 4 at the highest point. The open circle means that 4 is not included in the range.

STEP 3   Write the domain and range.

Domain: $\{x \mid -4 \le x < 3\}$
Range: $\{y \mid -1 \le y < 4\}$

Write the domain and range of each graph.

10.

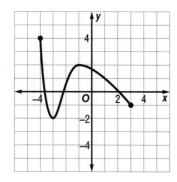

11.

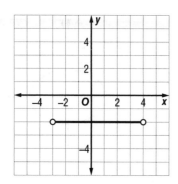

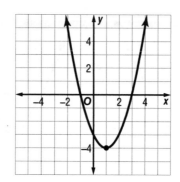

You can add, subtract, multiply, and divide functions just as you would with expressions. Consider the two functions $f(x)$ and $g(x)$. You can show operations with these functions as follows:

Addition	Subtraction	Multiplication	Division
$(f + g)(x)$ or $f(x) + g(x)$	$(f - g)(x)$ or $f(x) - g(x)$	$(fg)(x)$	$\frac{f}{g}(x)$

EXAMPLE 7 Given that $f(x) = 4x + 3$ and $g(x) = -2x - 4$, find $(f + g)(x)$ and $(f - g)(x)$.

To find $(f + g)(x)$, add the expressions.

$$f(x) + g(x) = (4x + 3) + (-2x - 4)$$
$$= 2x - 1$$

To find $(f - g)(x)$, subtract the expressions. Be sure to use parentheses.

$$f(x) - g(x) = (4x + 3) - (-2x - 4)$$
$$= 4x + 3 + 2x + 4$$
$$= 6x + 7$$

Therefore, $(f + g)(x) = 2x - 1$ and $(f - g)(x) = 6x + 7$.

The next example shows how to multiply and divide functions. Notice that you must list any exclusions from the solution set when you divide because you cannot divide by 0.

EXAMPLE 8 Given that $g(x) = x^2 - 4$ and $h(x) = x + 2$, find $(gh)(x)$ and $\left(\dfrac{g}{h}\right)(x)$.

To find $(gh)(x)$, multiply the expressions.

Use FOIL to multiply two binomials.

$$g(x) \cdot h(x)$$
$$(x^2 - 4)(x + 2)$$
$$= x^3 + 2x^2 - 4x - 8$$

To find $\left(\dfrac{g}{h}\right)(x)$, divide the expressions.
Use factoring to simplify each expression.
Then cancel. Show excluded values for x.

$$\left(\dfrac{g}{h}\right)(x)$$
$$\dfrac{x^2 - 4}{x + 2} = \dfrac{(x + 2)(x - 2)}{x + 2} = x - 2; x \neq -2$$

Therefore, $(gh)(x) = x^3 + 2x^2 - 4x - 8$ and $\left(\dfrac{g}{h}\right)(x) = x - 2; x \neq -2$.

Perform the operations as directed.

12. $f(x) = x^2 + x$
$g(x) = 4x - 4$
Find $(f + g)(x)$.

$g(x) = 2x - 3$
$h(x) = 3x - 4$
Find $(g + h)(x)$.

$f(x) = x - 3$
$h(x) = -x - 2$
Find $(f + h)(x)$.

13. $g(x) = x^3 + 2x$
$h(x) = 2x - 3$
Find $(g - h)(x)$.

$f(x) = 3x$
$g(x) = x - 4$
Find $(f - g)(x)$.

$f(x) = x^2 + 3$
$h(x) = -4$
Find $(f - h)(x)$.

14. $f(x) = 2x$
$g(x) = x + 4$
Find $(fg)(x)$.

$f(x) = x^2 - 1$
$h(x) = 3x - 1$
Find $(fh)(x)$.

$g(x) = 4x - 3$
$h(x) = x + 1$
Find $(gh)(x)$.

15. $f(x) = x^3 + x$
$g(x) = 2x$
Find $\left(\dfrac{f}{g}\right)(x)$.

$g(x) = x^2 - 16$
$h(x) = x + 4$
Find $\left(\dfrac{g}{h}\right)(x)$.

$f(x) = x + 3$
$h(x) = x^2 - x - 12$
Find $\left(\dfrac{f}{h}\right)(x)$.

Take a look at what happens when you substitute one function into another. This important process is called **composition.**

Take another look at the function machine. In a composition, instead of putting a single value of x into the machine, you put in an entire function.

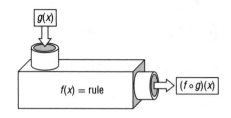

$(f \circ g)(x)$ means $f(g(x))$.
$(f \circ g)(x)$ is read "f of g of x."

EXAMPLE 9 Given that $f(x) = 4x - 1$ and $g(x) = 3x^2 + 2$, find $(f \circ g)(x)$.

Substitute g into f.	$f \circ g(x) = 4(3x^2 + 2) - 1$
Substitute $g(x)$ for x in $f(x)$.	$= 12x^2 + 8 - 1$
Simplify.	$= 12x^2 + 7$

Given that $f(x) = 3x$, $g(x) = x - 5$, and $h(x) = x^2$, find the compositions as directed.

16. $(f \circ g)(x) =$ $(h \circ g)(x) =$ $(f \circ f)(x) =$

      (**Hint:** Substitute g into f.)   (**Hint:** Substitute g into h.)   (**Hint:** Substitute f into itself.)

17. $(h \circ f)(x) =$ $(f \circ h)(x) =$ $(g \circ g)(x) =$

      (**Hint:** Substitute f into h.)   (**Hint:** Substitute h into f.)   (**Hint:** Substitute g into itself.)

Given that $f(x) = x^2 + 1$, $g(x) = 2x - 3$, and $h(x) = x - 7$, find each composition.

18. $(f \circ g)(x) =$ $(f \circ h)(x) =$ $(g \circ f)(x) =$

19. $(g \circ h)(x) =$ $(h \circ f)(x) =$ $(h \circ g)(x) =$

20. $(f \circ f)(x) =$ $(g \circ g)(x) =$ $(h \circ h)(x) =$

Graphing Linear and Quadratic Inequalities

In this lesson, you will learn to

- Graph inequalities on the coordinate plane.

The graph of a linear equation has three parts: the line, the half-plane above the line, and the half-plane below the line. In a linear equation, the solution is the line.

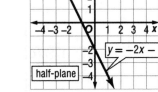

In a linear inequality, the solution is one of the half-planes and possibly the boundary line itself.

To graph inequalities, you need to be able to graph linear equations. You can review these steps on pages 65–66.

When you graph the boundary line, use a dashed line if the inequality symbol is < or >. The dashed line means that the line is not included in the solution set. Use a solid line if the inequality symbol is ≤ or ≥. The solid line means that the line is included in the solution set.

EXAMPLE 1   Graph: $y > 3x - 2$

STEP 1   Graph the equation $y = 3x - 2$. Draw a dashed line because the inequality uses the symbol >.

STEP 2   Choose a test point from one of the half-planes. Substitute it into the inequality. Try the origin (0, 0).

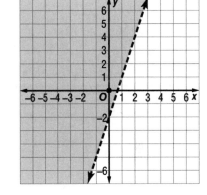

$y > 3x - 2$
$0 > 3(0) - 2$
$0 > -2$ True

The test point makes the inequality true. Shade the half-plane containing the test point.

You can use any convenient test point that does not lie on the boundary line.

EXAMPLE 2   Graph: $y \leq -x$

STEP 1   Graph the equation. Draw a solid line because the inequality uses the symbol ≤.

STEP 2   Choose a test point. Since the line passes through the origin, do not use (0, 0). Try (2, 2).

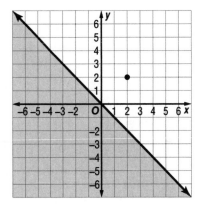

$y \leq -x$
$2 \leq -2$ False

The test point makes the inequality false. Shade the other side of the boundary line.

In Example 2, every point in the shaded part and on the boundary line makes the inequality true. The line is part of the graph. In Example 1, the line defines the boundary of the graph, but it is not included in the solution.

Graph.

1. $y > -3x - 1$ $y \le -2x - 2$

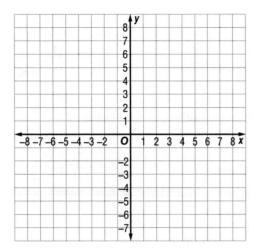

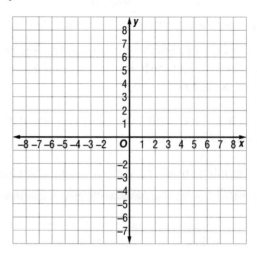

2. $y \ge \frac{2}{5}x - 1$ $y < \frac{1}{2}x + 5$

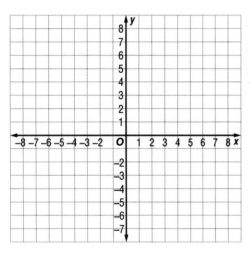

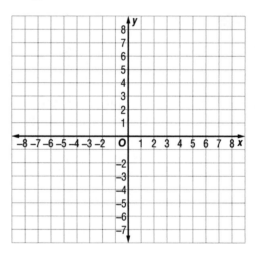

3. $y \ge \frac{5}{3}x$ $y < \frac{3}{4}x - 3$

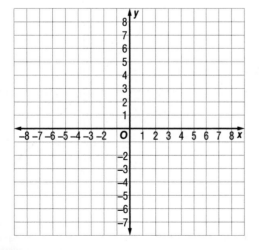

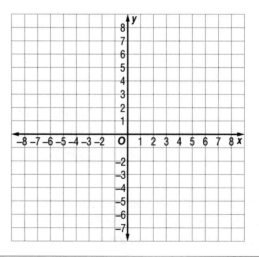

In a system of inequalities, the solution set is the set of values that makes both inequalities true. You can solve a system of inequalities by graphing both inequalities and shading the section of the graph that makes both inequalities true.

EXAMPLE 3   Graph: $y \geq 2x - 3$
$\qquad\qquad\quad y > -x + 3$

STEP 1   Graph both boundary lines.

STEP 2   Test a point in each section. The test point at (2, 3) will make both inequalities true.

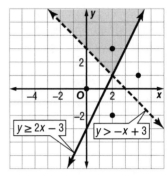

$$y \geq 2x - 3 \qquad\qquad y > -x + 3$$
$$3 \geq 2(2) - 3 \qquad\quad 3 > -(2) + 3$$
$$3 \geq 1 \quad \text{True} \qquad\quad 3 > 1 \quad \text{True}$$

STEP 3   Shade the section of the graph containing the point (2, 3).

Graph.

4. $y \leq -3x + 2$
$\quad y \geq x - 2$

$y \leq 2x + 3$
$y > -x - 3$

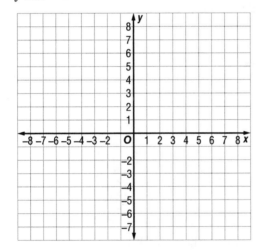

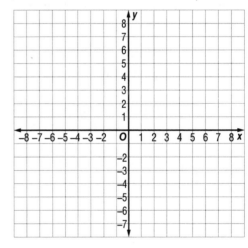

5. $y < -2x - 3$
$\quad y \leq x + 3$

$y > -x + 2$
$y > 3x - 2$

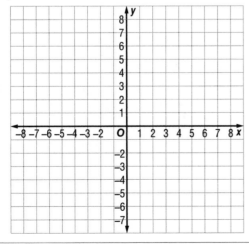

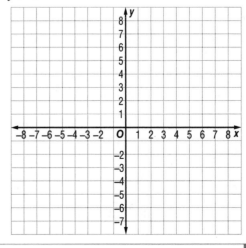

Trigonometric Ratios

In this lesson, you will learn to

- Find the sine, cosine, and tangent ratios of angles.
- Solve basic right triangle problems.

Look at ∠A in the diagram at the right. If you draw a line segment perpendicular to one of the legs of ∠A, a right triangle is formed.

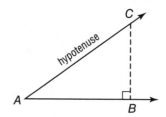

You know from your work with the Pythagorean theorem that the side which is opposite the right angle is the **hypotenuse.** The hypotenuse is always the longest side in a right triangle.

Think of the remaining legs in terms of ∠A. The leg that is part of ∠A is called the adjacent leg. The side farthest from point A is the opposite leg. The opposite leg does not connect to point A.

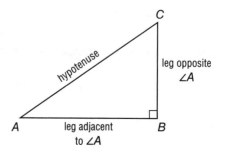

Mathematicians have discovered that the sides of a right triangle form special ratios. These ratios are the basis for our understanding of **trigonometry.** They are always written in terms of a particular angle.

sine of ∠A = $\dfrac{\text{length of leg opposite } \angle A}{\text{length of hypotenuse}}$
(sin A)

cosine of ∠A = $\dfrac{\text{length of leg adjacent to } \angle A}{\text{length of hypotenuse}}$
(cos A)

tangent of ∠A = $\dfrac{\text{length of leg opposite } \angle A}{\text{length of leg adjacent to } \angle A}$
(tan A)

 Math Talk Many students remember these ratios by learning the word:

SOH-CAH-TOA

The letters represent the words in each ratio:

- **S**ine-**O**pposite-**H**ypotenuse
- **C**osine-**A**djacent-**H**ypotenuse
- **T**angent-**O**pposite-**A**djacent

EXAMPLE 1 Find the sine, cosine, and tangent of ∠P.

STEP 1 Label the sides as they relate to ∠P. *PR* is the hypotenuse. *RQ* is opposite ∠P, and *PQ* is adjacent to ∠P.

STEP 2 Use the labels to write and simplify the ratios.

$\sin P = \dfrac{opp}{hyp} = \dfrac{12}{20} = \dfrac{3}{5}$

$\cos P = \dfrac{adj}{hyp} = \dfrac{16}{20} = \dfrac{4}{5}$

$\tan P = \dfrac{opp}{adj} = \dfrac{12}{16} = \dfrac{3}{4}$

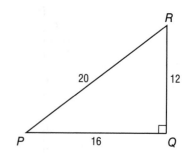

The ratios change if a different angle in the triangle is used.

EXAMPLE 2   Find the sine, cosine, and tangent of ∠R on page 214.

STEP 1   Label the sides as they relate to ∠R. *PR* is still the hypotenuse. *RQ* is adjacent to ∠R, and *PQ* is opposite ∠R.

STEP 2   Write and simplify the ratios.

$$\sin R = \frac{opp}{hyp} = \frac{16}{20} = \frac{4}{5} \quad \cos R = \frac{adj}{hyp} = \frac{12}{20} = \frac{3}{5} \quad \tan R = \frac{opp}{adj} = \frac{16}{12} = \frac{4}{3}$$

Find the sine, cosine, and tangent of the indicated angle.

1.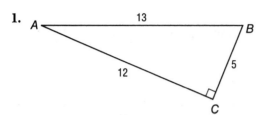

$\sin A =$ $\qquad$ $\sin B =$

$\cos A =$ $\qquad$ $\cos B =$

$\tan A =$ $\qquad$ $\tan B =$

2.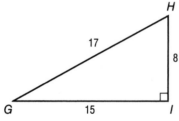

$\sin E =$ $\qquad$ $\sin F =$

$\cos E =$ $\qquad$ $\cos F =$

$\tan E =$ $\qquad$ $\tan F =$

3.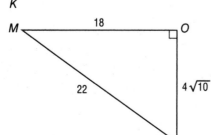

$\sin G =$ $\qquad$ $\sin H =$

$\cos G =$ $\qquad$ $\cos H =$

$\tan G =$ $\qquad$ $\tan H =$

4.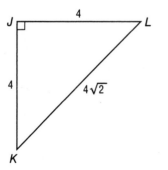

$\sin K =$ $\qquad$ $\sin L =$

$\cos K =$ $\qquad$ $\cos L =$

$\tan K =$ $\qquad$ $\tan L =$

5.

$\sin M =$ $\qquad$ $\sin N =$

$\cos M =$ $\qquad$ $\cos N =$

$\tan M =$ $\qquad$ $\tan N =$

Think about what you know about similar triangles. Two triangles are similar when their corresponding sides are proportional. In the similar triangles shown here, the angles are the same, but the sides of $\triangle DEF$ are twice the length of the sides of $\triangle ABC$.

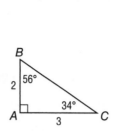
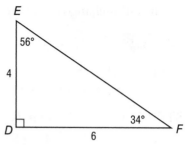

Look at $\angle B$ and $\angle E$. The sine, cosine, and tangent ratios for these angles must be the same. In fact, the trigonometric ratios for all angles measuring 56° must be the same. Therefore, the ratios for any angle measure remain constant.

Scientific calculators are programmed to display sine, cosine, and tangent ratios for any angle. You will need to have access to a scientific calculator with SIN, COS, and TAN keys for the remainder of this lesson.

EXAMPLE 3 Use a calculator to find sin 60°, cos 40°, and tan 80°. Round your answers to four decimal places.

	STEP 1	**STEP 2** Round display to 4 places.
sin 60°	SIN 60 Enter	0.866025404 *rounds to* 0.8660
cos 40°	COS 40 Enter	0.766044443 *rounds to* 0.7660
tan 80°	TAN 80 Enter	5.67128182 *rounds to* 5.6713

Therefore, sin 60° ≈ **0.8660**, cos 40° ≈ **0.7660**, and tan 80° ≈ **5.6713**.

Use a calculator to find the values of the trigonometric ratios. Round your answers to four decimal places.

6. sin 25°   cos 75°   tan 9°

7. cos 45°   sin 80°   tan 48°

8. tan 70°   cos 83°   sin 10°

9. sin 71°   cos 39°   tan 15°

10. cos 81°   sin 4°   tan 78°

11. cos 36°   tan 21°   sin 42°

Using the inverse ratios of sine, cosine, and tangent, you can work backward to find an angle measure. The inverse functions are labeled $\sin^{-1}$, $\cos^{-1}$, and $\tan^{-1}$ on your scientific calculator. Most calculators use a second function or shift key to access the trigonometric inverses. In other words, $\boxed{\sin^{-1}} = \boxed{\text{SHIFT}}\boxed{\text{SIN}}$.

EXAMPLE 4 The cosine of $\angle A$ is 0.4695. What is the measure of $\angle A$?

STEP 1 Use your calculator to find $\boxed{\cos^{-1}}$ 0.4695.

STEP 2 Round the display to a whole number of degrees:
61.99815465 rounds to 62°.

The measure of $\angle A$ is **62°**.

This process allows you to find an angle measure given any two sides of a right triangle. Simply figure out which trigonometric ratio you can write using the given sides. Then use the inverse of the ratio to find the angle measure.

EXAMPLE 5 What is the measure of $\angle R$ to the nearest degree?

STEP 1 Write a ratio using the given information.
You know that the hypotenuse of $\triangle RST$
measures 20 units. Side ST, which measures
12 units, is opposite $\angle R$.

Write the sine ratio: $\dfrac{\text{opposite}}{\text{hypotenuse}} = \dfrac{12}{20} = \dfrac{3}{5}$.

STEP 2 You know that $\sin R = \dfrac{3}{5}$. Use the inverse
of sine to solve for the angle.

Press: $\boxed{\sin^{-1}}\frac{3}{5}$ and $\boxed{\text{Enter}}$

Round to the nearest whole degree.

The measure of $\angle R$ is **37°**.

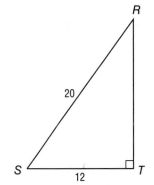

Find the angle measure using the information given. Round the angle measure to the nearest whole degree.

12. $\sin A = 0.9925$ $\cos B = 0.4226$ $\tan C = 2.7475$

  $m\angle A =$ $m\angle B =$ $m\angle C =$

13. $\sin D = 0.5592$ $\cos E = 0.9903$ $\tan F = 0.3640$

  $m\angle D =$ $m\angle E =$ $m\angle F =$

14. $\sin G = 0.3090$ $\cos H = 0.9135$ $\tan I = 2.1445$

  $m\angle G =$ $m\angle H =$ $m\angle I =$

15. $\sin J = 0.9511$ $\cos K = 0.8829$ $\tan L = 1.1504$

  $m\angle J =$ $m\angle K =$ $m\angle L =$

16.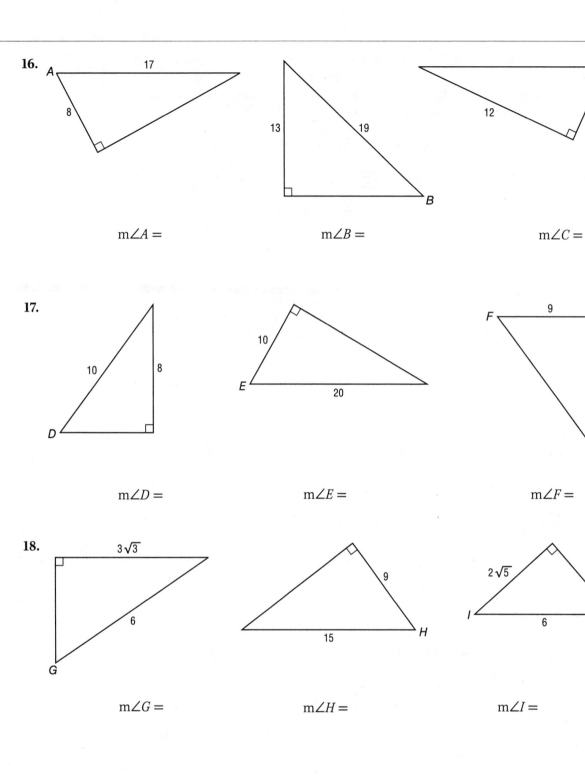

m∠A =   m∠B =   m∠C =

17.

m∠D =   m∠E =   m∠F =

18.

m∠G =   m∠H =   m∠I =

19.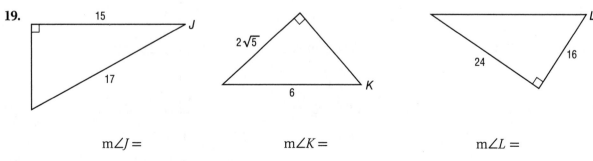

m∠J =   m∠K =   m∠L =

A right triangle is said to be "solved" when all of its side lengths and angle measures are known. There are several concepts and tools to help you solve right triangles.

- The sum of the interior angles of any triangle is 180º.
- The angles are related to the measures of the sides of a triangle using the ratios sine, cosine, and tangent.
- The Pythagorean theorem states that $c^2 = a^2 + b^2$, where c is the hypotenuse and a and b are the legs of a right triangle.

In the following example, angle measures are rounded to the nearest whole degree and side measures are rounded to the nearest tenth. Keep in mind that there is often more than one way to solve a triangle.

Use a calculator to find values of trigonometric ratios. In general, rounding these values to four decimal places will provide enough accuracy.

EXAMPLE 6   Solve △ABC. Round side measures to the nearest tenth.

STEP 1   Identify the given facts: ∠B measures 38º, and the side opposite ∠B measures 8 units.

STEP 2   Identify the missing parts. You need to find the hypotenuse, the side adjacent to ∠B, and the measure of ∠A.

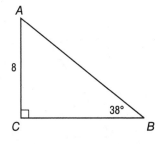

STEP 3   Find the missing parts one by one. Use the facts you have to solve for one variable at a time.

To solve for side AB, use the fact: $\sin = \dfrac{\text{opposite}}{\text{hypotenuse}}$.

Use a calculator to find the value of sin 38º.

Solve for side AB and round to the nearest tenth.

To solve for side BC, use the fact: $\tan = \dfrac{\text{opposite}}{\text{adjacent}}$.

Find tan 38º using a calculator.

Note that the Pythagorean theorem also could have been used to find the third side: $8^2 + b^2 = 13^2$.

$$\sin 38° = \frac{8}{AB}$$
$$0.6157 \approx \frac{8}{AB}$$
$$AB \approx \frac{8}{0.6157}$$
$$AB \approx 12.99 \approx \mathbf{13.0}$$

$$\tan 38° = \frac{8}{BC}$$
$$0.7813 \approx \frac{8}{BC}$$
$$BC \approx \frac{8}{0.7813}$$
$$BC \approx 10.23 \approx \mathbf{10.2}$$

All three side lengths are now known. To find the missing angle, remember that the sum of the interior angles of any triangle is 180º. Write an equation and solve for x.

$$90 + 38 + x = 180$$
$$128 + x = 180$$
$$x = 52$$

This triangle is solved.

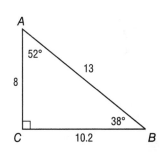

**Solve each triangle completely. Round side measures to the nearest tenth.
Round angle measures to the nearest whole degree.**

20.

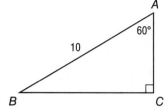

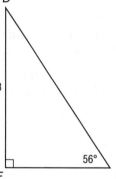

21.

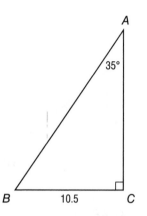

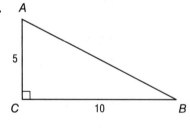

(**Hint:** $\cos E = \dfrac{\text{adjacent}}{\text{hypotenuse}} = \dfrac{2}{4}$

Solve for E using $\cos^{-1}$.)

22.

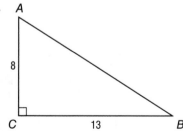

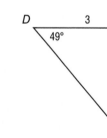

23.

Extension Topics Review

Solve the problems below. When you finish, check your answers at the back of the book and correct any errors.

Find the function values as indicated.

1. $f(x) = -4x - 2$ $\qquad$ $f(-9)$ $\qquad\qquad$ $f(5)$ $\qquad\qquad$ $f\left(\frac{1}{2}\right)$

2. $g(x) = -2x^3 + x$ $\qquad$ $g(-1)$ $\qquad\qquad$ $g(2)$ $\qquad\qquad$ $g(-3)$

3. $h(x) = x^2 - 5$ $\qquad$ $h(0)$ $\qquad\qquad$ $h(-3)$ $\qquad\qquad$ $h(4)$

4. $f(x) = -|-x + 2|$ $\qquad$ $f(6)$ $\qquad\qquad$ $f(-2)$ $\qquad\qquad$ $f(2)$

Write the domain of each function. Use set builder notation. Write $\mathbb{R}$ to indicate the set of all real numbers.

5. $f(x) = 2x - 7$ $\qquad$ $g(x) = |x^2 + 2|$ $\qquad$ $h(x) = 1 - \dfrac{1}{x - 3}$

6. $f(x) = \dfrac{x - 4}{x + 6}$ $\qquad$ $g(x) = \dfrac{1}{x^2 - 16}$ $\qquad$ $h(x) = \sqrt{3 - x}$

7. $f(x) = \dfrac{\sqrt{x - 3}}{x + 1}$ $\qquad$ $g(x) = \dfrac{2}{x^2 + 2x - 3}$ $\qquad$ $h(x) = \frac{2}{3}|x - 9|$

Determine whether each graph represents a function. Write *Yes* or *No*.

8.

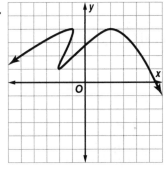 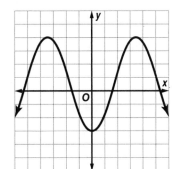 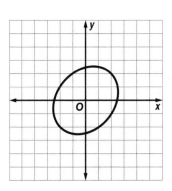

9.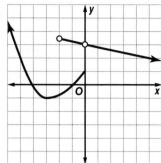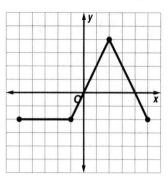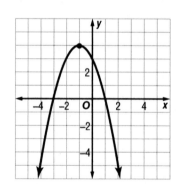

Write the domain and range of each graph.

10.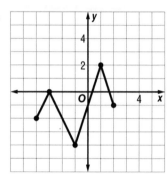

11.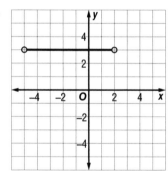

Perform the operations as directed. Indicate excluded values for x.

12. $f(x) = 3x + 5$ $f(x) = x - 3$ $g(x) = -x + 5$

   $g(x) = 4x - 4$ $h(x) = 3x + 1$ $h(x) = 2x - 2$

   Find $(f + g)(x)$.   Find $(fh)(x)$.   Find $(g - h)(x)$.

13. $h(x) = 3x - 5$ $f(x) = -2x + 4$ $f(x) = x - 2$

   $g(x) = -2x^2 - 1$ $g(x) = x^2 - 3$ $h(x) = x^2 - 2x$

   Find $(h - g)(x)$.   Find $(f + g)(x)$.   Find $\left(\dfrac{f}{h}\right)(x)$.

14. $f(x) = 3x + 5$
$g(x) = 2x$
Find $(fg)(x)$.

$f(x) = x + 5$
$h(x) = x^2 + 6x + 5$
Find $\left(\dfrac{f}{h}\right)(x)$.

$g(x) = x - 3$
$h(x) = 4x + 1$
Find $(g + h)(x)$.

15. $f(x) = x^3 - 5x^2$
$g(x) = 2x - 3$
Find $(f - g)(x)$.

$g(x) = -x + 1$
$h(x) = 3x + 5$
Find $(gh)(x)$.

$f(x) = 3x + 9$
$h(x) = x + 3$
Find $\left(\dfrac{f}{h}\right)(x)$.

Given that $f(x) = 2x + 5$, $g(x) = x^2 - 1$, and $h(x) = 3x - 2$, find each composition.

16. $(f \circ g)(x)$ $(g \circ h)(x)$ $(f \circ h)(x)$

17. $(g \circ f)(x)$ $(f \circ f)(x)$ $(g \circ g)(x)$

Given that $f(x) = 2x^2$, $g(x) = 4x + 5$, and $h(x) = 3x - 7$, find each composition.

18. $(h \circ f)(x)$ $(f \circ g)(x)$ $(h \circ g)(x)$

Graph each linear inequality.

19. $y \leq -x - 1$ $y < \dfrac{3}{2}x + 4$

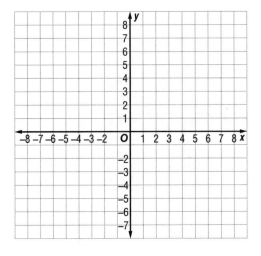 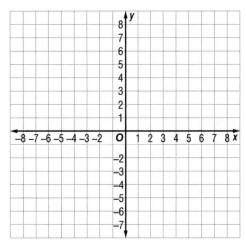

20. $y > \frac{3}{5}x - 1$ $\qquad\qquad\qquad\qquad$ $y \leq 3x - 5$

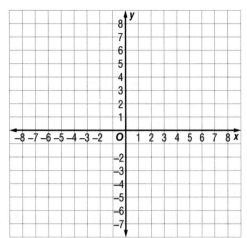 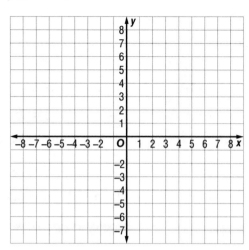

Graph each system of inequalities.

21. $y \leq -2x + 3$ $\qquad\qquad\qquad\qquad$ $y < -2x + 2$

$\quad\ y < 2x - 1$ $\qquad\qquad\qquad\qquad\ \ $ $y > -\frac{1}{2}x - 1$

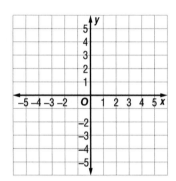 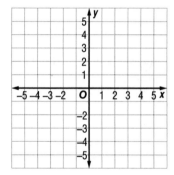

22. $y \leq 3$ $\qquad\qquad\qquad\qquad\qquad\ \ $ $y \leq x + 3$

$\quad\ y < -\frac{5}{2}x - 2$ $\qquad\qquad\qquad\qquad$ $y \geq -5x - 3$

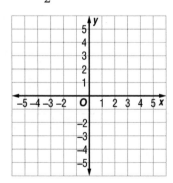 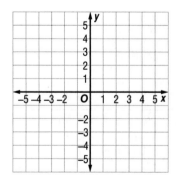

Find the sine, cosine, and tangent of the indicated angle.

23.

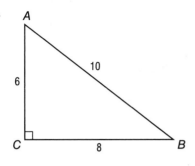

sin A =   sin B =

cos A =   cos B =

tan A =   tan B =

24.

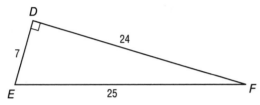

sin E =   sin F =

cos E =   cos F =

tan E =   tan F =

25.

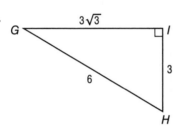

sin G =   sin H =

cos G =   cos H =

tan G =   tan H =

Using a calculator, find the angle measure from the information given.
Round the angle measure to the nearest whole degree.

26. $\sin A = 0.7193$ $\cos B = 0.9848$ $\tan C = 0.2867$

m∠A =   m∠B =   m∠C =

27. $\sin D = 0.4226$ $\cos E = 0.3746$ $\tan F = 2.2460$

m∠D =   m∠E =   m∠F =

28. $\sin G = 0.9945$ $\cos H = 0.7771$ $\tan I = 1.1918$

m∠G =   m∠H =   m∠I =

29.

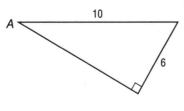

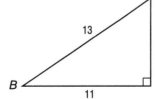

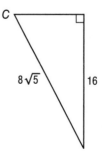

m∠A =   m∠B =   m∠C =

Solve each triangle completely. Round side measures to the nearest tenth.
Round angle measures to the nearest whole degree.

30.

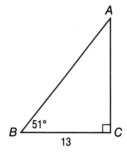

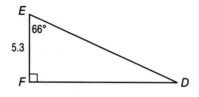

31.

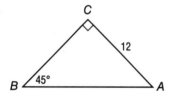

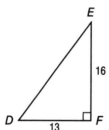

32.

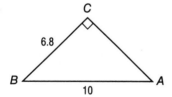

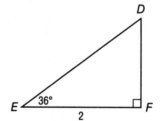

ANSWER KEY

Pages 3–4

1. $6 + 3, -2(9), 9 + (-7)$
2. $3 \cdot (5 \cdot 6), (-2 + 6) + 10, (5 \cdot -4) \cdot -9$
3. $72 - 108, -2x - 10, 27a - 3ab$
4. $4(3a + 2b), 7(3x - 1), 6(x + 2y)$
5. $5(2a - b + 3), -3(x + 2y), 2x(5 + y)$
6. $3x, 3, \frac{1}{12}$
7. $2, \frac{1}{8}, \frac{16b}{25}$
8. $27, 11, 67$
9. $56, 27, 9$
10. $6 \cdot (4 - 2) + 3 = 15$, True, $(7 - 4)^2 - 1 = 8$
11. $(4 + 2 \cdot 2)^2 = 64, 3^2 \cdot (8 - 7 + 2)^2 = 81$, True
12. $\frac{82 + 96 + 90 + 84}{4} = 88, \frac{7 + 7 + 4 + 7 + 15}{5} = \$8$
13. $\frac{16 + 25 + 19 + 21 + 31 + 14}{6} = 21$,
    $\frac{140 + 390 + 220}{3} = 250$

Pages 6–8

1. $32, 64, -9$
2. $1, 64, 14$
3. $1, 63, 26$
4. $3^7, x^8, (-y)^4$
5. $a^{10}, 2^{13}, (-b)^0 = 1$ and $1^3 = 1$
6. $a^8, 4^{12}, 12^{15}$
7. $9, \frac{9}{16}, \frac{1}{9}$
8. $\frac{1}{125}, 7, 16$
9. $25a^2, 27x^6, a^2 b^4 c^6$
10. $y^{10}, \frac{x^3}{y^6}, \frac{b^3}{16}$
11. $\frac{c^2}{b^3}, 2x, \frac{z^2}{x^6 y^4}$
12. $a^{-4}, \frac{1}{a^4}; y^{-3}, \frac{1}{y^3}; m^{-5}n^{-2}, \frac{1}{m^5 n^2}$
13. $\frac{n^2}{m^3}, \frac{1}{x^3 y^3}, \frac{2^5 b^3}{a^7}$
14. 5.0×10^5
15. 3.0×10^4
16. $4.5 \times 10^9, 6.8 \times 10^{-3}$
17. $4.2 \times 10^5, 2.0 \times 10^6$
18. 9.3×10^7 miles

Pages 9–12

1. $7, -12, 6, -9$
2. $2, 5, -4, -1$
3. $5\sqrt{2}, 3\sqrt{6}, 10\sqrt{3}$
4. $3\sqrt[3]{4}, 2\sqrt[3]{3}, 3\sqrt[3]{2}$
5. $12\sqrt{3}, 12\sqrt{5}, 10\sqrt{3}$
6. $24\sqrt{3}, 4\sqrt[3]{5}, 8\sqrt[3]{7}$
7. $\frac{1}{4}, \frac{2}{5}, \frac{7}{11}$
8. $\frac{\sqrt{3}}{10}, \frac{3}{4}, \sqrt{5}$
9. $\frac{\sqrt{6}}{3}, \frac{\sqrt{21}}{6}, \frac{\sqrt{6}}{4}$
10. $\frac{\sqrt{10}}{6}, \frac{\sqrt{6}}{10}, \frac{\sqrt{42}}{12}$
11. $xy, 6a^2 \sqrt{2a}, 4m^2 n$
12. $\frac{2n\sqrt{3}}{5}, \frac{a\sqrt{14}}{7}, \frac{m^2\sqrt{3m}}{3n}$
13. $\sqrt[6]{5^5}, \sqrt[12]{x^7}$
14. $4^2, m^{\frac{3}{2}} = m\sqrt{m}$

Pages 13–14, Rules and Properties Review

1. **b.** associative, **c.** distributive, **a.** commutative
2. $5(5a - 1), 4(10x + 3y), -9(m + 3n)$
3. $3y(7 + 2x), a(1 - 8b), -4(4m - n + 3)$
4. $\frac{y}{4}, -\frac{1}{20}, 4$
5. $-\frac{12}{n}, \frac{5}{6}, -\frac{3x}{2}$
6. **c.** 30, **b.** 22, **a.** 6
7. $13, 5, 64$
8. $\frac{17 + 23 + 30 + 16 + 24}{5} = 22$
9. $5x^3 y^4, \frac{8b^6 c^3}{a^6}, \frac{m^3}{9}$
10. $2x^3, \frac{4n}{m}, \frac{a^4}{b^2 c^6}$
11. $\frac{1}{x^3}, \frac{1}{2mn^6}, \frac{b^7 c^2}{a^6}$
12. $4.8 \times 10^{-2}, 3.2 \times 10^3$
13. $9\sqrt{5}, 8\sqrt{2}, 4\sqrt[3]{3}$
14. $\frac{3}{5}, \frac{10}{3}, \frac{\sqrt{15}}{9}$
15. $\sqrt{80x^6} = \sqrt{16 \cdot x^4 \cdot x^2 \cdot 5} = 4x^3\sqrt{5}$,
    $3\sqrt{72m^3} = 3\sqrt{36 \cdot m^2 \cdot 2m} = 18m\sqrt{2m}$
16. $x^{\frac{7}{10}} = \sqrt[10]{x^7}, a^3$

Page 16

1. $m - n, b + 5$
2. $3xy, \frac{a}{2} - 8$
3. $x^2 - xy, \frac{b - c}{b + c}$

4. $3(m - n), 4(x^2 + y)$

5. $(x + 3)(x - 3), 6(x - 7)$

6. $\frac{x - 5}{x^3}, 5x + x$

7. $\frac{x + 9}{2}$ or $\frac{1}{2}(x + 9), (x + 5)(x - 5)$

Pages 17–20

1. $9x + 2, -7a + 2b$

2. $15m + 5n - 13, 11x^2 + 5x - 15$

3. $-3x^3 - x^2, -3y - 1$

4. $-16c^3 + 8cd + 7d^3, 5m^2n - 8mn + 4mn^2$

5. $11a - 3, 3x^2 - 3$

6. $3n^3 + n^2 - 5n + 8, 3x^2 + 2x - 3$

7. $3c^3 + 2c^2 + c + 1, y - 9$

8. $-4n^2 - n, 6a^2 + 2a - 2$

9. $4x^2 + 4x - 6, 5m + 2n$

10. $5x^3y^2, -12a^5b^5, -6m^5n^3$

11. $2mn + 9n, 8a^2 + 24ab, -4xy + y^2$

12. $-11x^2 + 22xy, -21xy - 14y^2, 5m^2 - 30mn$

13. $-2m^2 - 14m + 10, 3a^3 + 15a^2 + 45a,$
$-2b^3 + 3b^2 + b$

14. $4x^3 - 12x^2 - 24x, -2y^3 - 2y^2 + 16y,$
$15n^3 + 35n^2 - 10n$

15. $m^2 + 7m + 12, a^2 - 7a + 10, 2c^2 - 11c - 6$

16. $m^2 - 25, y^2 + 10y + 16, -x^2 + 11x - 18$

17. $6y^2 - 2y - 8, a^2 - 6a + 5, 2x^2 - xy - 6y^2$

18. $x^2 + 8x + 16, n^2 - 10n + 25, -a^2 + 2a + 15$

Pages 21–24

1. $\frac{x}{8}, \frac{-3m}{n}$

2. $\frac{2c - 3}{c - 3}, 4$

3. $\frac{2}{m}, n - 3$

4. $\frac{3a + 5}{a^2}, \frac{3y - 2x}{xy}$

5. $\frac{8x + 15}{20x^3}, \frac{5 - 2n}{5n^2}$

6. $\frac{6y + 3x}{xy^2}, \frac{4a^2 - 2c^2}{abc}$

7. $\frac{8y + 3}{5(y + 1)}, \frac{-m + 5}{6(m + 1)}$

8. $\frac{10b - 3}{b(b - 1)}, \frac{2(9x - 1)}{3x(3x - 1)}$

9. $\frac{7}{3(c - 2)}, \frac{-21}{10(y + 1)}$

10. $\frac{9}{5(b - 2)}, \frac{-3}{n - 4}$

11. $\frac{-4}{a - 2}, \frac{3}{x - 7}$

12. $\frac{5a + 1}{(a + 5)(a - 1)}, \frac{9x + 8}{6(x - 3)}$

13. $\frac{3(y - 20)}{(y + 4)(y - 8)}, \frac{2(3a + 1)}{(a + 3)(a - 1)}$

14. $\frac{3c - 2}{(c - 1)(c + 1)}, \frac{5x + 7}{(x + 1)(x - 4)}$

15. $\frac{3a + 1}{(a - 1)(a - 2)}, \frac{-3x^2 - 4x - 10}{(x + 2)(x + 3)(x - 5)}$

Pages 26–27

1. $-3, 29, -9$

2. $-10, 15, 25$

3. $10, 2, 8, 12$

4. $35, -11, -1, 2$

5. $7, 4, -9, -10$

6. **a.** 44 in., **b.** 121 sq in.

7. **a.** 56 cm, **b.** 192 sq cm

8. **a.** 60 units, **b.** 150 sq units

9. **a.** 51 units, **b.** 96 sq units

Pages 28–30, Variables Review

1. $13 + 5m, y - x$

2. $\frac{10}{n}, 3(P - Q)$

3. $\frac{b + c}{bc}, m^2 - n$

4. $2(x + 8), \frac{5x}{x^3}$

5. $7x + 3, -4m + 7n$

6. $-10a^2 + 9, 4y^2 - 3y + 3$

7. $x^3 + 11, 2a - 12$

8. $4n^2 - n - 1, -x^2 - 6x + 9$

9. $-3n^2 - n - 12, 2a - 9b$

10. $4a^4bc^3, 6x^2 + 12xy, -42n - 35n^2$

11. $12a^2 + 4ab, 2x^4 - 2x^3y, d^2 + d - 2$

12. $5a^2 + 12a - 9, -m^2 + 5m + 14, 3x^3y^2 + 3x^2y^3$

13. $b^2 + b - 12, 2d^2 - d - 15, x^2 - 8x + 16$

14. $-m^2 + m + 6, -x^2 + 12x - 35, n^2 + 6n + 9$

15. $\frac{6x - 5}{4}, \frac{-2y - 7}{y - 2}$

16. $\frac{5y - 2x}{xy}, \frac{-4a + 3}{a^2}$

17. $\frac{5n + 1}{n(n + 1)}, \frac{y - 2}{y - 1}$

18. $\frac{1}{m + 3}, -\frac{2}{(x + 2)(x - 3)}$

19. $-36, 2, 3$

20. $40, -6, 147$

21. $-27, 10, 2$

22. $1, 39, 8$

23. a. 46 ft, **b.** 120 sq ft

24. a. 56 m, **b.** 196 sq m

25. a. 42 in., **b.** 84 sq in.

26. a. 40 cm, **b.** 60 sq cm

13. 0, all real numbers, $\varnothing$

14. $\varnothing$, 0, all real numbers

15. all real numbers, $\varnothing$, all real numbers

16. $\frac{1}{5}, -\frac{2}{3}$

17. $0.6, 0.1$

18. $-\frac{8}{3}, -1$

19. $-5, -0.6$

20. $\frac{3}{2}, -1.1$

21. $16, 3$

Pages 32–33

1. $x = -2$

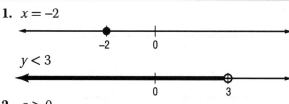

$y < 3$

2. $a > 0$

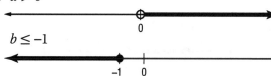

$b \leq -1$

3. $n = 4$

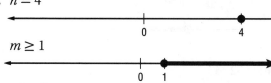

$m \geq 1$

4. $c > -3$

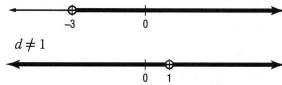

$d \neq 1$

5. True, False

6. False, True

7. True, False

8. False, True

9. True, True

10. True, False

Page 35

1. $6a = 54, x = -5$

2. $y = -12, -14 = 2b$

3. $-4n = -5n + 15, x = \frac{18}{x}$

4. $2x = 3 + x, -a = 7$

5. $3 = 8m, 21 = 7y$

6. $x + 9 = 10, 3a + 1 = 7$

Pages 36–41

1. $3, 10, -3$

2. $-6, -1, 3$

3. $4, -1, 2$

4. $-6, -7, 5$

5. $-5, 15, -8$

6. $-30, 18, 20$

7. $6, -4$

8. $-1, -2$

9. $0, 7$

10. $8, -5$

11. $12, 6$

12. $1, -9$

Page 43

1. 31

2. 75

3. 29 and 34

4. 105

5. -7 and -5

6. 52

7. 10 years old

8. -3 and -2

9. 24 points

10. 3rd test score: 93

Page 45

1. $117°$

2. $16°$

3. $96°$

4. $113°$

5. 9 cm, 9 cm, and 4 cm

6. $144°$

7. 17 in.

8. 9 cm and 4 cm

Page 47

1. 65 tickets

2. 12 years old

3. 15 years old

4. 36 quarters

5. 13 hours

6. 230 miles

7. 18 3-point questions

8. 22 points

9. 11 quarters

10. $\frac{7}{11}$

Pages 49–51

1. $n < -2$

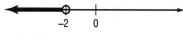

$x \leq -1$

$a > -4$

2. $y \geq 3$

$m < 2$

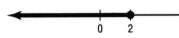

$n < 0$

3. $b \leq 2$

$x > 0$

$y \leq 8$

4. $x < 0$

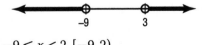

$a \leq -2$

$y > -1$

5. $0 \leq x < 3, [0, 3)$

$n \geq 1$ or $n \leq -7, (-\infty, -7] \cup [1, \infty)$

6. $-5 \leq b < -3, [-5, -3)$

$-8 \leq x < 0, [-8, 0)$

7. $m > 3$ or $m < -9, (-\infty, -9) \cup (3, \infty)$

$-9 \leq x < 2, [-9, 2)$

8. $b \geq 6$ or $b < 5, (-\infty, 5) \cup [6, \infty)$

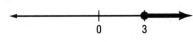

$2 < x < 3, (2, 3)$

9. $a < 1$ or $a \geq 7, (-\infty, 1) \cup [7, \infty)$

$y \leq -4$ or $y > 0, (-\infty, -4] \cup (0, \infty)$

10. $-2 \leq x < 1, [-2, 1)$

$1 < m \leq 3, (1, 3]$

Pages 52–55

1. $8, 1, -5$

2. $18, 10, 12$

3. $0, 16, -6$

4. $b = -\frac{1}{5}, 3; n = \frac{9}{2}, 0;$ no solution

5. $x = 7, 1; a = 13, -4; y = 1, 3$

6. $c = -2; n = 10, 2; x = 6, -8$

7. $-2 \leq n \leq 2, [-2, 2]$

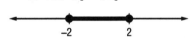

$a < 0$ or $a > 10, (-\infty, 0) \cup (10, \infty)$

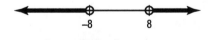

8. $x < -8$ or $x > 8, (-\infty, -8) \cup (8, \infty)$

$-4 < b < 0, (-4, 0)$

9. no solution, $\emptyset$

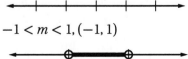

$-1 < m < 1, (-1, 1)$

10. $-8 \le x \le -2, [-8, -2]$

a = all real numbers, $(-\infty, \infty)$

11. $y < -5$ or $y > 1, (-\infty, -5) \cup (1, \infty)$

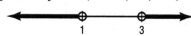

b = all real numbers, $(-\infty, \infty)$

12. $-2 < x < 4, (-2, 4)$

$y < 1$ or $y > 3, (-\infty, 1) \cup (3, \infty)$

Pages 57–58

1. $w = \frac{p - 2l}{2}, y = \frac{c - ax}{b}$

2. $x = -\frac{b}{a}, b = \frac{3V}{h}$

3. $r = \frac{A - P}{Pt}, b_1 = \frac{2A}{h} - b_2$

4. $C = \frac{5}{9}F - \frac{160}{9}, R = \frac{100D}{100 - x}$

5. $R = S - \frac{A}{T}, h = \frac{3V}{\pi r^2}$

6. $p = \frac{CT}{V}, F = \frac{9}{5}C + 32$

7. $n = \frac{D}{180} + 2, h = t - Ia$

8. $y = -\frac{2}{5}x + 4, y = -\frac{1}{2}x + \frac{3}{8}$

9. $y = \frac{3}{2}x, y = \frac{1}{4}x + 4$

10. $y = 5x + 2, y = 3x$

11. $y = 6x - 2, y = \sqrt{x^2 - 2}$

Pages 59–64, Equation and Inequality Basics Review

1.

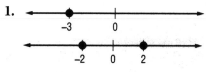

2.

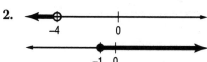

3.

4. True, False

5. False, True

6. False, True

7. True, False

8. False, True

9. $5a - 10 = -a, -3b + 12 = -6b$

10. $3x - 5 > 1, 2 < 2n < 8$

11. $y < 3, |5 - m| \ge 2$

12. $x = 4, a = -6, y = -3$

13. $n = -3, y = 0, x = -2$

14. $b = -5, m = 1, n = -1$

15. $m = -27, a = 16, x = 0$

16. $b = -\frac{1}{6}, n = \frac{3}{5}, y = -\frac{7}{6}$

17. $a = 4, y = 1, n = -8$

18. C is correct.

The incorrect line in A is $-4x = -7$.

The incorrect line in B is $-3 - 2x = -10 - x$.

19. 94 points

20. -16 and -14

21. 28 dimes

22. $26°$

23. 9 inches

24. 12 large plants and 18 medium plants

25. $120°$

26. $85

27. 70 spaces

28. $90°, 30°$, and $60°$

29. $n \le -4$

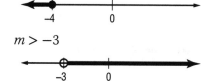

$m > -3$

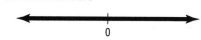

all real numbers

30. $x \geq 2$

$y \geq -2$

$b \geq 0$

31. $-5 < x < -1, (-5, -1)$

$a < -8$ or $a \geq 2, (-\infty, -8) \cup [2, \infty)$

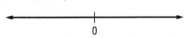

32. $0 \leq b \leq 1, [0, 1]$

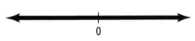

no solution, $\varnothing$

33. all real numbers

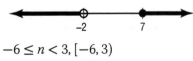

$n < -4$ or $n \geq 8, (-\infty, -4) \cup [8, \infty)$

34. $2 \leq a \leq 3, [2, 3]$

no solution, $\varnothing$

35. $m < -2$ or $m \geq 7, (-\infty, -2) \cup [7, \infty)$

$-6 \leq n < 3, [-6, 3)$

36. $3, 8, 24$

37. $15, 0, -12$

38. $n = -15, 7; a = 6, 8; x = -5, 5$

39. $x = -2, 20; b = -15, 5; n = 0$

40. $b = -11, 4; x = -3, \frac{1}{2}; a = 0, \frac{1}{2}$

41. $y = -3, -2; n = 0, 1; m = -\frac{9}{5}, 5$

42. $x = \frac{9}{2}; m = 0, -18; a = \varnothing$

43. $n = 4, 1; x = 3, -5; y = -2$

44. $b < -7$ or $b > -1, (-\infty, -7) \cup (-1, \infty)$

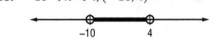

$x < -4$ or $x > 4, (-\infty, -4) \cup (4, \infty)$

45. $-10 < n < 4, (-10, 4)$

no solution, $\varnothing$

46. $-9 < x < 7, (-9, 7)$

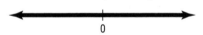

all real numbers, $(-\infty, \infty)$

47. $-6 < n < 7, (-6, 7)$

all real numbers, $(-\infty, \infty)$

48. $b = \frac{2A}{h}, p = \frac{I}{rt}$

49. $N = \frac{2.5H}{d^2}, h = \frac{3V}{b}$

50. $c = 8h - 4R, b = 3A - (a + c)$

51. $b = \frac{1 - da}{d}, t = \frac{3k}{V}$

52. $h = \frac{s - 2lw}{2l + 2w}, b_2 = \frac{2A}{h} - b_1$

53. $y = x + 2, y = 2x - 3$

54. $y = \frac{3}{4}x, y = \frac{1}{8}x + 1$

Pages 65–68

1. $D(-3, 6)$	**3.** $F(2, 1)$
2. $E(-4, 0)$	**4.** $G(7, -5)$

5–10.

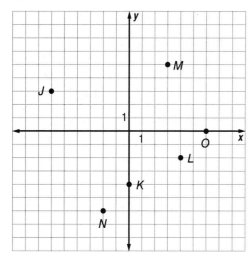

18.

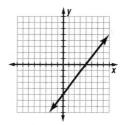

19. Answers will vary.

Pages 70–74

1. $(1, 0), (0, 3)$

3. $(-3, 0), (0, -2)$

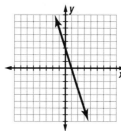

2. $(5, 0), (0, 4)$

4. $(2, 0), (0, 4)$

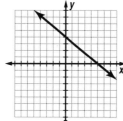

11.

x	y
1	3
2	4
3	5

x	y
-3	-5
0	-4
3	-3

12.

x	y
-4	-3
0	0
4	3

x	y
-4	2
-2	-1
0	-4

13.

x	y
-2	12
0	2
2	-8

x	y
4	-3
8	-4
12	-5

5. $\frac{4}{5}, -\frac{5}{2}$

7. $\frac{9}{5}, -\frac{4}{7}$

6. $-\frac{1}{2}$, undefined

8. $0, -\frac{1}{4}$

9.

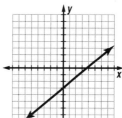

10.

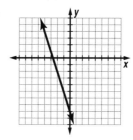

14.

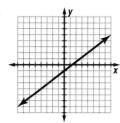

16.

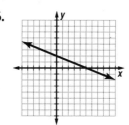

15.

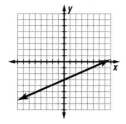

17.

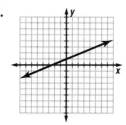

11. $\frac{1}{5}, -\frac{1}{2}$

13. $\frac{5}{4}$, undefined

12. $4, 0$

14. $0, -2$

Pages 77–82

1. **a.** $y = -3x + 4$
   b. $3x + y = 4$

2. **a.** $y = x - 3$
   b. $x - y = 3$

3. **a.** $y = 2x - 5$
   b. $2x - y = 5$

4. **a.** $y = -x$
   b. $x + y = 0$

5. **a.** $y = \frac{2}{3}x + 3$
   b. $2x - 3y = -9$

6. **a.** $y = -\frac{2}{5}x + 2$
   b. $2x + 5y = 10$

7. **a.** $y = \frac{7}{4}x - 1$
   b. $7x - 4y = 4$

8. **a.** $y = -\frac{3}{4}x - 2$
   b. $3x + 4y = -8$

9. $y = x - 3$

10. $y = -\frac{3}{2}x - 1$

11. $y = \frac{4}{5}x$

12. $y = -4x + 4$

13. $y = -8x + 5, y = 3x + 1$

14. $y = -7x + 2, y = 4x - 1$

15. $y = \frac{1}{2}x + 5, y = -2$

16. $y = \frac{1}{4}x - 4, y = \frac{6}{5}x + 5$

17. $y = -\frac{10}{3}x + \frac{5}{3}, y = \frac{1}{2}x + \frac{3}{2}$

18. $2x - 5y = 15, 9x - 5y = 25$

19. $7x + 4y = -16, 7x - 3y = -15$

20. $5x - 2y = -10, 2x + 3y = 0$

21. $x + 3y = -12, 2x + 5y = 10$

22. $5x - 3y = 3, x - 5y = -20$

23. $y = -\frac{5}{2}x + 5, y = \frac{1}{2}x + \frac{1}{2}$

24. $y = -\frac{3}{4}x + 1, y = \frac{1}{3}x - \frac{7}{3}$

25. $y = -\frac{7}{6}x - \frac{1}{2}, y = -\frac{10}{3}x + 5$

26. $y = x, y = \frac{3}{8}x - \frac{3}{2}$

27. $x + y = 0$

28. $x + 6y = 8$

29. $2x - y = 7$

30. $3x - 4y = -16$

31. $2x + 3y = 9$

32. $x - 4y = 3$

Pages 84–86

1. parallel, neither, perpendicular

2. perpendicular, parallel, neither

3. perpendicular, neither, parallel

4. $y = -x + 2, y = 6x + 8$

5. $y = -3x + 2, y = \frac{6}{5}x - 1$

6. $y = \frac{1}{2}x - 1, y = \frac{3}{2}x + 5$

7. $y = -x + 2, y = x$

8. $y = \frac{3}{4}x - \frac{7}{2}, y = \frac{5}{3}x - \frac{20}{3}$

9. $6x - y = 2, x + y = 4$

10. $5x - 3y = -9, x - 4y = -9$

11. $AB: y = -2x - 10$
    $AC: y = \frac{1}{2}x + 5$
    $CD: y = -2x + 15$
    $BD: y = \frac{1}{2}x - \frac{5}{2}$

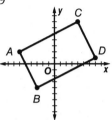

12. No. The slope of AD is $-\frac{1}{13}$. The slope of BC is $\frac{11}{7}$. Because one slope is not the negative reciprocal of the other, the diagonals cannot be perpendicular.

Pages 88–92

1. $(-4, 6), (-7, -3)$

2. $(6, 3), (8, 3)$

3. $(-5\frac{1}{2}, -1), (0, -\frac{1}{2})$

4. $(6, 0), (-8\frac{1}{2}, 1\frac{1}{2})$

5. midpoint of side $PQ = (-2\frac{1}{2}, -2)$
   midpoint of side $QR = (2\frac{1}{2}, -1)$
   midpoint of side $PR = (1, 4)$

6. 5

8. $6\sqrt{2}$

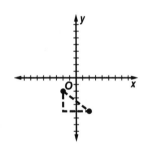

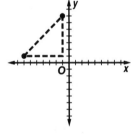

7. 9

9. $3\sqrt{10}$

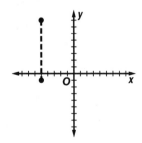

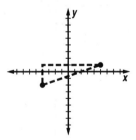

10. $4\sqrt{5}, 12$

11. $11\sqrt{2}, 6\sqrt{2}$

12. $3\sqrt{13}, 3\sqrt{2}$

13. $9, 4\sqrt{5}$

14. $5\sqrt{2}, 4\sqrt{10}$

15. Yes, side $AB =$ side $AC = 5\sqrt{10}$.

16. $14\sqrt{5}$

Pages 93–98, Linear Equations and Graphing Review

1.

x	y
1	7
2	9
3	11

x	y
−3	−4
0	−2
3	0

2. $(2, 0), (0, 5)$ $(4, 0), (0, −3)$

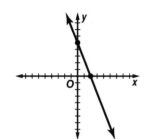 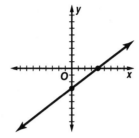

3. $−3, \frac{1}{2}$

4.

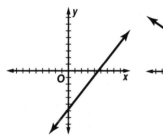 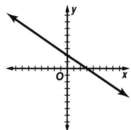

5.

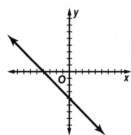 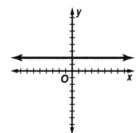

6. $4, -\frac{7}{3}$

7. $\frac{2}{5}$, undefined

8. $-\frac{1}{2}, \frac{3}{10}$

9. $−2, 0$

10. $6x − y = 5, x − 4y = 12$

11. $2x − 3y = −9, 3x + 4y = 0$

12. $3x − 2y = 4, x − 5y = 15$

13. $y = x + 4, y = -\frac{1}{4}x + 2$

14. $y = -\frac{5}{3}x − 2, y = −5$

15. $x = 3, y = -\frac{5}{6}x - \frac{7}{3}$

16. $x + 4y = 20, 3x + 2y = 6$

17. $5x − 2y = 8, 7x − 2y = −6$

18. $4x + 3y = −8, 2x − 7y = 22$

19. $4x − 3y = −14, 2x + y = −11$

20. $x − y = −2, 3x + 7y = −9$

21. $x = −3, x + 5y = 1$

22. $x − 2y = 8, 8x + y = 11$

23. parallel, neither, perpendicular

24. neither, parallel, perpendicular

25. $y = −x + 7, y = \frac{1}{6}x + \frac{1}{3}$

26. $x = −2, y = \frac{3}{5}x + 2$

27. $y = -\frac{5}{4}x - \frac{7}{4}, y = 3x + 2$

28. $y = \frac{7}{4}x − 3, y = x$

29. $(1, −4), (−6, 2)$

30. $(0, 6\frac{1}{2}), (−3, 2\frac{1}{2})$

31. $(7\frac{1}{2}, 3), (5, 2)$

32. $10, 4\sqrt{2}$

33. $2\sqrt{10}, 5$

34. $9\sqrt{2}, 4\sqrt{5}$

35. $PQ: \sqrt{5}$ $QR: 2\sqrt{10}$
     $RS: 5\sqrt{5}$ $SP: 2\sqrt{10}$

36. Yes. Sides PQ and SR both have the same slope, 2.

37. midpoint of AB: $(−3, 4)$
midpoint of CD: $(0, −5)$
distance: $3\sqrt{10}$

Pages 100–101

1. $(1, −3)$ $(5, 1)$

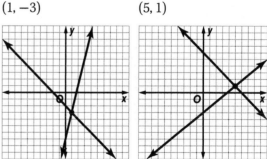

2. $(−8, 6)$ $(4, 2)$

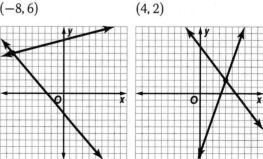

3. $(-1, 2)$ $(-3, -6)$

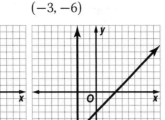

4. $(3, -3)$ $(-2, -5)$

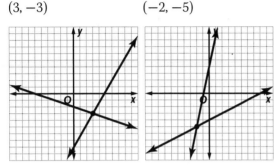

5. $(4, -5)$ $(-2, -4)$

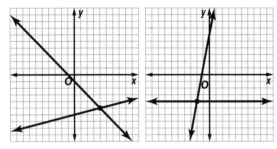

Pages 102–109

1. $(3, 0), (-1, 6), (-5, 5)$
2. $(3, -1), (-1, 2), (-4, -5)$
3. $(7, 5), (-1, -1), (-2, -1)$
4. $(8, 10), (-4, 0), (6, -4)$
5. $(1, 8), (7, -7), (3, -6)$
6. $(5, -8), (-10, 3), (4, -3)$
7. $(-4, -4), (2, 1), (0, -3)$
8. inconsistent, $(6, -3), (-2, -3)$
9. $(0, 0), (-4, 0), (-9, 2)$
10. $(-5, -2), (2, -2)$, dependent
11. dependent, $(-2, 4)$, inconsistent
12. inconsistent, dependent, $(-1, 0)$
13. $(0, 1), (-3, -1), (-1, 5)$

14. $(6, 2), (9, -7), (-1, 9)$
15. $(7, -4), (-1, 0), (3, 4)$
16. $(-1, 4), (2, 3), (1, 2)$
17. $(5, 3), (-9, -4), (1, 0)$
18. $(-3, 6), (0, -9), (6, -1)$
19. inconsistent, $(1, 8), (5, 6)$
20. dependent, $(6, -3), (9, 7)$
21. $(0, 0), (-4, 2)$, dependent
22. $(1, -1), (4, 6), (-4, 3)$
23. $(5, 2)$, inconsistent, dependent
24. $(2, -1), (-5, 8), (-3, -2)$
25. $(4, 0, -3), (-4, 2, 0)$

Pages 111–113

1. 6 in.   10. 21
2. 85 dimes   11. 15
3. 7 years old   12. 74
4. 15 and 36   13. 93
5. 41 compact spaces   14. 18
6. 85 points   15. $\frac{2}{7}$
7. 40 student tickets   16. $\frac{13}{17}$
8. 30 lb   17. $\frac{7}{13}$
9. 58

Pages 114–116

1. 9 inches   7. $y = 8x, y = 0.25x$
2. 60 nickels   8. $y = \frac{45}{x}, y = \frac{32}{x}$
3. 30 blue marbles   9. $y = 2.5x, y = \frac{28}{x}$
4. 225 lunch customers   10. $y = \frac{800}{x}, y = 20x$
5. 36 students   11. $y = 16x, y = 8x$
6. 200 customers   12. $y = \frac{36}{x}, y = \frac{288}{x}$

Pages 118–122

1. 10 oz of A and 5 oz of B
2. 1 qt of 90% and 9 qt of 70%
3. 3 ft³ of 40% and 6 ft³ of pure clay
4. 30 mL of 65% and 20 mL of 40%

5. 4 kg

6. 3 L

7. 37%

8. 46%

9. 50%

10. 4 kg

11. 6 fl oz

12. 7 kg

13. $5\frac{5}{6}$ hours

14. 24 minutes

15. 2 hours

16. 36 minutes

17. 30 hours

18. 20 hours

19. 5 hours

20. 7 hours

21. 4 hours

22. $7\frac{1}{5}$ hours

23. 6 minutes

24. 4 hours

Pages 125–127

1. 360 km

2. 60 mph

3. 18 km/h

4. 20 miles

5. 7.5 mph

6. 214 km/h

7. boat: 27 mph, current: 6 mph

8. plane: 170 mph, wind: 30 mph

9. 7 mph

10. jet: 180 mph, wind: 20 mph

11. $\frac{3}{4}$ hour downstream, $4\frac{1}{2}$ miles

12. 6 km/h

Pages 129–132

1. length = 36 cm, width = 30 cm

2. 120 cm

3. 6 in.

4. 400 sq ft

5. 48 yd

6. width = 5 in., length = 11 in.

7. 12 in., 12 in., and 18 in.

8. 12 sq units

9. 145

10. a. $x = 4, y = 3$
    b. $AE = 18$ cm

11. a. 48 in.
    b. 38.4 in.

Pages 133–138, Problem Solving with Multiple Variables Review

1. $(7, -5)$ $(1, 6)$

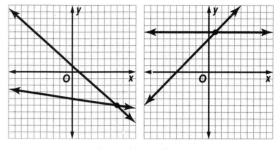

2. $(-4, -2)$ $(-4, 2)$

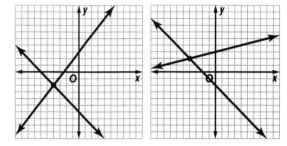

3. $(3, -5)$ $(8, 6)$

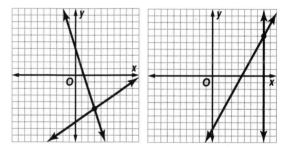

4. $(2, -7), (1, -3), (-3, -8)$

5. $(-5, -1), (4, 8)$, inconsistent

6. $(-3, 4)$, dependent, $(0, 5)$

7. $(-1, -2), (9, 2), (-6, 8)$

8. $(-2, 2)$, dependent, $(3, 1)$

9. inconsistent, $(5, -6), (-4, 3)$

10. The student's mistake is in Step B. The student only worked with the 2nd equation. The value of x should have been substituted into the 1st equation. The correct answer is $(1, 4)$.

11. 1,530 children's tickets

12. 39

13. $8

14. 11 years old

15. $\frac{13}{20}$

16. 18

17. $y = 4x, y = 1.6x$

18. $y = \frac{27}{x}, y = \frac{28}{x}$

19. $y = 3.2x, y = \frac{90}{x}$

20. 18 green and 42 white marbles

21. 36 and 20

22. 35 dimes and 20 nickels

23. 120 customers

24. 8 m^3 of soil A and 3 m^3 of soil B

25. 8 oz of Sun Valley and 24 oz of Rancho

26. 3 drops

27. 30% alcohol

28. 75% of the resulting alloy

29. 4 oz of 88% iron and 12 oz of pure iron

30. 30 minutes

31. 12 hours

32. 36 hours

33. 10 hours

34. $3\frac{3}{4}$ days

35. $11\frac{1}{4}$ hours

36. 50 km/h

37. 6 hours

38. boat: 27 mph, current: 9 mph

39. 5 mph

40. 168 cm

41. length = 40 in., width = 18 in.

42. 10 cm, 10 cm, and 15 cm

43. 112 yards

44. 8 in.

45. length = 15 m, width = 6 m

46. 135 sq in.

47. 9 square units

Page 141

1. $7x^2 - 5 = 0; a = 7, b = 0, c = -5$
$x^2 - 4x - 12 = 0; a = 1, b = -4, c = -12$
$3x^2 - 20 = 0; a = 3, b = 0, c = -20$

2. $x^2 + 8x - 48 = 0; a = 1, b = 8, c = -48$
$4x^2 - 8 = 0; a = 4, b = 0, c = -8$
$x^2 - 12x + 11 = 0; a = 1, b = -12, c = 11$

3. $x^2 - 8x - 18 = 0; a = 1, b = -8, c = -18$
$2x^2 + x - 120 = 0; a = 2, b = 1, c = -120$
$x^2 + 7x - 11 = 0; a = 1, b = 7, c = -11$

4. $3x^2 - 3x - 5 = 0; a = 3, b = -3, c = -5$
$11x^2 + x - 24 = 0; a = 11, b = 1, c = -24$
$3x^2 - 8 = 0; a = 3, b = 0, c = -8$

Pages 143–149

1. $\{-7, -1\}$, $\{6, -3\}$, $\{-4, 4\}$

2. $\{-2, 4\}$, $\{-5\}$, $\left\{-1, \frac{4}{3}\right\}$

3. $\{6, 0\}$, $\left\{\frac{1}{7}, 5\right\}$, $\left\{\frac{5}{6}, 2\right\}$

4. $\left\{-\frac{1}{4}, 0\right\}$, $\left\{-\frac{1}{6}, -\frac{3}{8}\right\}$, $\left\{\frac{5}{2}, \frac{4}{5}\right\}$

5. $\{10\}$, $\{-1, 3\}$, $\left\{-\frac{3}{8}, \frac{3}{2}\right\}$

6. $\{3, 8\}$, $\{6, -8\}$, $\{2, 1\}$

7. $\{-7, 9\}$, $\{-4, -7\}$, $\{-6, 4\}$

8. $\{-7, 10\}$, $\{-5, -10\}$, $\{-1, -7\}$

9. $\{7, -5\}$, $\{-9, 8\}$, $\{9, 3\}$

10. $\{-6, -5\}$, $\{4, -2\}$, $\{-7, -11\}$

11. $\{9, -5\}$, $\{-5, 6\}$, $\{9, 6\}$

12. $\{10, -3\}$, $\{5, 9\}$, $\{6, -1\}$

13. $\{3, -4\}$, $\{-12, 7\}$, $\{-6, -12\}$

14. $\{-3, -5\}$, $\{11, 6\}$, $\{9, -6\}$

15. $\{0, -4\}$, $\{0, 3\}$, $\left\{0, -\frac{9}{5}\right\}$

16. $\{0, -10\}$, $\left\{0, \frac{2}{3}\right\}$, $\left\{0, \frac{5}{4}\right\}$

17. $\left\{0, \frac{5}{2}\right\}$, $\left\{0, \frac{2}{9}\right\}$, $\left\{0, -\frac{2}{3}\right\}$

18. $\left\{\frac{7}{3}, 7\right\}$, $\left\{-\frac{9}{2}, 1\right\}$, $\left\{\frac{1}{3}, 5\right\}$

19. $\left\{\frac{9}{7}, 6\right\}$, $\left\{-\frac{9}{7}, 1\right\}$, $\left\{\frac{9}{2}, 2\right\}$

20. $\left\{\frac{11}{3}, -10\right\}$, $\left\{-\frac{7}{2}, -10\right\}$, $\left\{-\frac{3}{5}, -12\right\}$

21. $\left\{-\frac{1}{5}, -8\right\}$, $\left\{\frac{2}{5}, 5\right\}$, $\left\{-\frac{7}{3}, -8\right\}$

22. $\left\{-\frac{2}{3}, 1\right\}$, $\left\{-\frac{7}{4}, 7\right\}$, $\left\{\frac{1}{10}, -1\right\}$

23. $\left\{-\frac{8}{5}, \frac{1}{2}\right\}$, $\left\{-\frac{3}{2}, -\frac{2}{11}\right\}$, $\left\{\frac{9}{5}, -\frac{6}{5}\right\}$

24. $\left\{\frac{4}{3}, -5\right\}$, $\left\{-\frac{5}{11}, 9\right\}$, $\left\{-\frac{9}{8}, 1\right\}$

25. $\left\{\frac{7}{3}, 1\right\}$, $\left\{\frac{5}{2}, -\frac{10}{11}\right\}$, $\left\{\frac{5}{3}, -\frac{1}{2}\right\}$

26. $\left\{\frac{9}{7}, 1\right\}$, $\left\{-\frac{5}{8}, 12\right\}$, $\left\{\frac{3}{4}, -9\right\}$

Page 152

1. $(n - 12)(n + 12)$, $(4x - 1)(4x + 1)$, $(9a - 11)(9a + 11)$

2. $(c^2 + 4)(c - 2)(c + 2)$, $(9n^2 + 1)(3n - 1)(3n + 1)$, $(x - 6)^2$

3. $(x - 8)^2$, $(5a + 1)^2$, $(4x - 3)^2$

4. $\{-8, 8\}$, $\left\{-\frac{1}{4}, \frac{1}{4}\right\}$, $\left\{-\frac{11}{7}, \frac{11}{7}\right\}$

5. $\left\{-\frac{7}{10}, \frac{7}{10}\right\}$, $\{12\}$, $\left\{\frac{2}{3}\right\}$

6. $\{-8\}$, $\left\{\frac{5}{6}\right\}$, $\left\{-\frac{7}{8}\right\}$

Page 155

1. $(x^2 + 2)(6x + 1)$, $(4a^2 + 7)(a + 7)$, $(3m^2 + 2)(4m - 3)$

2. $(b^2 + 1)(9b - 8)$, $(5x^2 - 4)(x + 3)$, $(n^2 - 10)(4n + 3)$

3. $(3x + 1)(3x - 1)(2x + 1)$, cannot be factored, $(9n^2 + 7)(2n + 1)$

4. $(6a^2 - 1)(8a + 5)$, $2(4x^2 - 7)(x + 2)$, cannot be factored

5. $(x + 2)(x - 2)(7x + 5)$, $(2a^2 - 5)(9a - 1)$, $(2m + 1)(2m - 1)(8m + 5)$

6. $(a + 1)(a - 1)(4a - 1)$, $(2x^2 + 5)(3x + 1)$, $(b^2 - 2)(4b + 1)$

7. cannot be factored, $4(3b^2 + 2)(2b - 1)$, $(5m^2 + 4)(m + 5)$

8. $2(2n^2 - 1)(5n + 3)$, cannot be factored, $(4a^2 + 3)(a - 2)$

9. $3(4m^2 - 3)(5m - 2)$, $(n + 2)(n - 2)(4n + 3)$, $(b + 1)(b - 1)(5b - 2)$

Page 157

1. $\frac{3x - 1}{2(x + 1)}$, $\{-5, -1\}$; $\frac{7a + 1}{2(a + 5)}$, $\{0, -5\}$; $\frac{2x - 1}{3x + 4}$, $\left\{-\frac{4}{3}, 9\right\}$

2. $\frac{5n}{3(n - 2)}$, $\{-7, 2\}$; $\frac{2x + 5}{5(x - 2)}$, $\{-4, 2\}$; $\frac{3b + 2}{9b}$, $\{0, -1\}$

3. $\frac{8}{3a(2a - 5)}$, $\left\{0, \frac{5}{2}, -1\right\}$; $\frac{n(2n - 5)}{n - 2}$, $\{2, 10\}$; $\frac{4n}{3}$, $\{-3, 1\}$

4. $\frac{4}{3x(x + 1)}$, $\{0, -1, -5\}$; $\frac{3(5b + 1)}{10(b - 2)}$, $\{0, 2, 6\}$; $\frac{(2x - 9)(x + 2)}{2x - 5}$, $\left\{\frac{5}{2}\right\}$

5. $\frac{3x - 2}{2(x + 2)(x - 2)}$, $\{2, -2\}$; $\frac{3}{2m(m + 3)}$, $\{0, -3, 6\}$; $2(n + 1)$, $\{5\}$

6. $\frac{2a}{3}$, $\left\{-\frac{2}{5}, -1\right\}$; $\frac{2n - 1}{2n(n - 1)}$, $\{0, 1\}$; $\frac{10(x - 1)}{3(3x - 4)}$, $\left\{\frac{4}{3}, 3\right\}$

Pages 158–162

1. $8\sqrt{7}$, $3\sqrt{6}$, $\sqrt{10}$

2. $-2\sqrt{2}$, $34\sqrt{5}$, $-3\sqrt{2}$

3. $13\sqrt{2}$, 0, $-13\sqrt{6}$

4. $36\sqrt{2}$, $-2\sqrt{10}$, $21\sqrt{5}$

5. $16\sqrt{2}$, $-26\sqrt{3}$, $-42\sqrt{2}$

6. $2a\sqrt{3a}$, $60x^2\sqrt{x}$, $50m^2$

7. $12\sqrt{2b} + 2\sqrt{6b}$, $-12 - 4\sqrt{30x}$, $2n\sqrt{3} + 5\sqrt{6n}$

8. $5\sqrt{3x} + 6\sqrt{10}$, $-12\sqrt{10n} - 40n$, $-20\sqrt{b} + 5b^2\sqrt{5}$

9. $4 + 8\sqrt{3}$, -23, $18 + 4\sqrt{15}$

10. $3x + 2\sqrt{3x} - 15$, $15 - 4\sqrt{10a} - 8a$, $3m$

11. $\frac{3\sqrt{5}}{10}$ $\frac{\sqrt{2}}{2}$ $\frac{\sqrt{6m}}{m^2}$

12. $\frac{\sqrt{15}}{10}$ $\frac{2\sqrt{2}}{n^2}$ $\frac{3\sqrt{15b}}{5}$

13. $\frac{-\sqrt{5} - \sqrt{10x}}{25x^2}$ $\frac{4a\sqrt{3} - \sqrt{6}}{18}$ $\frac{2m\sqrt{m} - 4\sqrt{m}}{m^2}$ or $\frac{2\sqrt{m}(m - 2)}{m^2}$

14. $\frac{-\sqrt{2} - 6n}{10n}$ $\frac{\sqrt{3} + 3\sqrt{x}}{2x}$ $\frac{-\sqrt{2} - 2b\sqrt{2}}{3b}$

15. $\frac{\sqrt{5a} - a^2}{5 - a^3}$ $\frac{4 + \sqrt{2x}}{8 - x}$ $\frac{15 - 5\sqrt{2m}}{9 - 2m}$

16. $\frac{-10 - 2b\sqrt{b}}{25 - b^3}$ $\frac{3n\sqrt{2n} - 2n\sqrt{6}}{6n - 8}$ $\frac{2\sqrt{3a} - 2a\sqrt{15}}{3 - 15a}$

17. $\frac{-5x^2 + \sqrt{2x}}{25x^4 - 2x}$ $\frac{-10 + 25a\sqrt{3a}}{4 - 75a^3}$ $\frac{8 - 2\sqrt{2}}{7m^2}$

Pages 164–168

1. $\{2, -12\}$, $\{8, -10\}$, no solution

2. $\{3, 5\}$, no solution, $\{3, 13\}$

3. $\{-9\}$, $\left\{-\frac{1}{2}, -\frac{5}{2}\right\}$, $\left\{\frac{8}{3}, -2\right\}$

4. no solution, $\{2, 3\}$, $\left\{\frac{7}{6}, -\frac{5}{6}\right\}$

5. $\{5 + 3\sqrt{2}, 5 - 3\sqrt{2}\}$, $\{2 + 2\sqrt{6}, 2 - 2\sqrt{6}\}$, $\{1 + \sqrt{10}, 1 - \sqrt{10}\}$

6. $\{-1, -4\}$, $\{4 + \sqrt{11}, 4 - \sqrt{11}\}$, $\{3 + \sqrt{15}, 3 - \sqrt{15}\}$

7. $\left\{\frac{7}{2}, -\frac{3}{2}\right\}$, $\{1 + 2\sqrt{3}, 1 - 2\sqrt{3}\}$, $\{3, -1\}$

8. $\{2 + \sqrt{10}, 2 - \sqrt{10}\}$, $\left\{\frac{1}{2}, -\frac{5}{2}\right\}$, $\{4, -2\}$

9. $\left\{3, -\frac{17}{8}\right\}$, $\left\{\frac{14}{5}, -4\right\}$, $\left\{\frac{8}{3}, -\frac{5}{2}\right\}$

10. $\left\{\frac{-8 + \sqrt{145}}{3}, \frac{-8 - \sqrt{145}}{3}\right\}$, $\left\{-\frac{1}{2}, -\frac{4}{3}\right\}$, $\left\{\frac{-1 + \sqrt{21}}{4}, \frac{-1 - \sqrt{21}}{4}\right\}$

11. $\left\{\dfrac{-8+2\sqrt{7}}{3}, \dfrac{-8-2\sqrt{7}}{3}\right\}$, $\left\{\dfrac{7+2\sqrt{21}}{5}, \dfrac{7-2\sqrt{21}}{5}\right\}$,

$\left\{\dfrac{-3+3\sqrt{13}}{2}, \dfrac{-3-3\sqrt{13}}{2}\right\}$

12. $\left\{\dfrac{8+\sqrt{55}}{3}, \dfrac{8-\sqrt{55}}{3}\right\}$, $\left\{\dfrac{-1+2\sqrt{5}}{2}, \dfrac{-1-2\sqrt{5}}{2}\right\}$,

$\left\{\dfrac{1+5\sqrt{13}}{6}, \dfrac{1-5\sqrt{13}}{6}\right\}$

Pages 171–172

1. $\left\{8, -\dfrac{9}{2}\right\}$, $\{1, -5\}$, $\left\{\dfrac{9}{4}, -4\right\}$

2. $\left\{\dfrac{14}{3}, -3\right\}$, $\left\{\dfrac{2+\sqrt{19}}{5}, \dfrac{2-\sqrt{19}}{5}\right\}$, $\left\{\dfrac{\sqrt{10}}{2}, -\dfrac{\sqrt{10}}{2}\right\}$

3. $\{-4+2\sqrt{7}, -4-2\sqrt{7}\}$, $\left\{\dfrac{3+\sqrt{85}}{2}, \dfrac{3-\sqrt{85}}{2}\right\}$, $\{-3, -4\}$

4. $\left\{4, -\dfrac{2}{3}\right\}$, $\left\{\dfrac{\sqrt{2}}{2}, -\dfrac{\sqrt{2}}{2}\right\}$, $\left\{\dfrac{1+3\sqrt{5}}{2}, \dfrac{1-3\sqrt{5}}{2}\right\}$

5. $\left\{\dfrac{4\sqrt{3}}{3}, -\dfrac{4\sqrt{3}}{3}\right\}$, $\left\{-\dfrac{5}{4}\right\}$, $\left\{\dfrac{1+\sqrt{21}}{10}, \dfrac{1-\sqrt{21}}{10}\right\}$

6. $\{5+\sqrt{29}, 5-\sqrt{29}\}$, $\left\{\dfrac{5+4\sqrt{2}}{7}, \dfrac{5-4\sqrt{2}}{7}\right\}$, $\left\{\dfrac{5}{2}\right\}$

7. $\left\{\dfrac{\sqrt{14}}{2}, -\dfrac{\sqrt{14}}{2}\right\}$, $\left\{\dfrac{15}{4}, -3\right\}$, $\left\{\dfrac{3+2\sqrt{15}}{3}, \dfrac{3-2\sqrt{15}}{3}\right\}$

8. two roots, $\left\{\dfrac{9}{5}, -3\right\}$; no solution; one root, $\{2\}$

9. no solution; two roots, $\left\{\dfrac{5+\sqrt{11}}{2}, \dfrac{5-\sqrt{11}}{2}\right\}$;

two roots, $\left\{\dfrac{13}{4}, -4\right\}$

10. one root, $\left\{\dfrac{7}{3}\right\}$; no solution;

two roots, $\left\{\dfrac{2+2\sqrt{11}}{5}, \dfrac{2-2\sqrt{11}}{5}\right\}$

Page 175

1. 81 sq in.

2. 15

3. 60 cm

4. about 4 sec

5. 18 mph

6. 6 mph

7. about 3 sec

8. 20 ft by 28 ft

Pages 176–180, Quadratic Equations Review

1. $4x^2 + 16x - 9 = 0$, $a = 4$, $b = 16$, $c = -9$;

$5n^2 - n - 21 = 0$, $a = 5$, $b = -1$, $c = -21$;

$6a^2 + 13 = 0$, $a = 6$, $b = 0$, $c = 13$

2. $\left\{0, \dfrac{8}{7}\right\}$, $\{9, 4\}$, $\{5, -7\}$

3. $\{3\}$, $\left\{1, -\dfrac{7}{10}\right\}$, $\left\{\dfrac{7}{2}, -\dfrac{2}{3}\right\}$

4. $(x+2)(x-10)$, $\{10, -2\}$;

$3(a-1)(a+2)$, $\{1, -2\}$;

$(m-9)(m-6)$, $\{9, 6\}$

5. $(b+2)(b+9)$, $\{-2, -9\}$;

$2x(7x+3)$, $\left\{0, -\dfrac{3}{7}\right\}$;

$(7n-2)(n+10)$, $\left\{\dfrac{2}{7}, -10\right\}$

6. $(5a-4)(a+1)$, $\left\{\dfrac{4}{5}, -1\right\}$;

$(2m+9)(m-4)$, $\left\{-\dfrac{9}{2}, 4\right\}$;

$2(3x-4)(x-3)$, $\left\{\dfrac{4}{3}, 3\right\}$

7. $(4x-1)(2x+7)$, $\left\{\dfrac{1}{4}, -\dfrac{7}{2}\right\}$;

$(3m-2)(3m+1)$, $\left\{\dfrac{2}{3}, -\dfrac{1}{3}\right\}$;

$(a-7)(10a-9)$, $\left\{7, \dfrac{9}{10}\right\}$

8. $5(b+1)(9b+5)$, $\left\{-1, -\dfrac{5}{9}\right\}$;

$(n-8)(8n-9)$, $\left\{8, \dfrac{9}{8}\right\}$;

$4(2x+7)(3x-2)$, $\left\{-\dfrac{7}{2}, \dfrac{2}{3}\right\}$

9. $(n+8)(n-8)$, $(5x+2)(5x-2)$, $(7c+10)(7c-10)$

10. $(4a-1)^2$, $(x-11)^2$, $(9a^2+4)(3a+2)(3a-2)$

11. $(n+7)^2$, $(5b-9)^2$, $8m(4m+3)^2$

12. $(2x^2-3)(5x+8)$, $(2m^2+1)(2m-7)$, $5(a^2+4)(a-2)$

13. $(2b^2+9)(b-4)$, $(4x^2+5)(3x-1)$, $(n+2)(n-2)(5n-4)$

14. $3(7n^2+1)(4n+9)$, $(5a^2-7)(9a+1)$, $3x(x^2+7)(8x+3)$

15. $\dfrac{1}{x+4}$, $\{-4, 2\}$; $n-6$, $\{2\}$; $\dfrac{5}{2(a+4)}$, $\{0, -4\}$

16. $\dfrac{b-1}{9}$, $\{-8\}$; $\dfrac{4}{n+9}$, $\{-9, 1\}$; $\dfrac{x-1}{x-5}$, $\{-1, 5\}$

17. $\dfrac{1}{3}$, $\{0, 2, 6\}$; $\dfrac{n+7}{8(n+2)}$, $\{-2, 3\}$; $\dfrac{a(a-5)}{2(a-1)}$, $\{1, 8\}$

18. $-5\sqrt{3}$, $5\sqrt{6}$, 0

19. $3\sqrt{3}$, $2\sqrt{2}$, $-6\sqrt{2}$

20. $-13\sqrt{2}$, $-13\sqrt{7}$, $-4\sqrt{2}$

21. $\dfrac{4\sqrt{10m}}{5}$, $\dfrac{3\sqrt{10}}{5n}$, $\dfrac{2x\sqrt{3x}}{3}$

22. $\dfrac{3\sqrt{7a}}{7}$, $\dfrac{2\sqrt{10x}+3x\sqrt{5}}{5x}$, $\dfrac{5\sqrt{2}-2\sqrt{10}}{16b}$

23. $\dfrac{-16+4\sqrt{5}}{11}$, $\dfrac{15-5\sqrt{3a}}{9-3a}$, $\dfrac{2\sqrt{3}+2}{m}$

24. $\{2, -10\}$, $\{8, -12\}$, no solution

25. $\{7, 3\}$, no solution, $\{2, 0\}$

26. no solution, $\left\{-\frac{1}{2}, -\frac{5}{2}\right\}$, $\{-3, -4\}$

27. $\{7 + 4\sqrt{5}, 7 - 4\sqrt{5}\}$, $\left\{\frac{2 + 3\sqrt{11}}{2}, \frac{2 - 3\sqrt{11}}{2}\right\}$, $\left\{\frac{5}{2}, -\frac{9}{2}\right\}$

28. $\left\{\frac{-3 + 3\sqrt{17}}{8}, \frac{-3 - 3\sqrt{17}}{8}\right\}$, $\left\{3, -\frac{15}{8}\right\}$, $\left\{-\frac{1}{2}, -\frac{6}{5}\right\}$

29. $\{-3 + \sqrt{10}, -3 - \sqrt{10}\}$, $\{1 + 2\sqrt{2}, 1 - 2\sqrt{2}\}$, $\{4 + 2\sqrt{10}, 4 - 2\sqrt{10}\}$

30. $\left\{\frac{3}{2}, -\frac{5}{3}\right\}$, $\left\{\frac{-3 + \sqrt{33}}{4}, \frac{-3 - \sqrt{33}}{4}\right\}$, $\left\{\frac{17 + \sqrt{65}}{2}, \frac{17 - \sqrt{65}}{2}\right\}$

31. $\left\{1, -\frac{7}{3}\right\}$, $\left\{\frac{3 + 3\sqrt{5}}{2}, \frac{3 - 3\sqrt{5}}{2}\right\}$, $\left\{\frac{7 + \sqrt{29}}{10}, \frac{7 - \sqrt{29}}{10}\right\}$

32. $\left\{\frac{-1 + \sqrt{5}}{4}, \frac{-1 - \sqrt{5}}{4}\right\}$, $\left\{\frac{\sqrt{2}}{2}, \frac{-\sqrt{2}}{2}\right\}$, $\left\{2, \frac{2}{3}\right\}$

33. $\left\{\frac{-4 + 4\sqrt{6}}{5}, \frac{-4 - 4\sqrt{6}}{5}\right\}$, $\{2 + \sqrt{10}, 2 - \sqrt{10}\}$, $\left\{\frac{-4 + \sqrt{13}}{3}, \frac{-4 - \sqrt{13}}{3}\right\}$

34. $\left\{\frac{-7 + \sqrt{5}}{2}, \frac{-7 - \sqrt{5}}{2}\right\}$, $\left\{\frac{1 + \sqrt{11}}{4}, \frac{1 - \sqrt{11}}{4}\right\}$, $\{-3 + \sqrt{7}, -3 - \sqrt{7}\}$

35. two roots, $\{-4, -3\}$; no solution; one root, $\{0\}$

36. no solution; one root, $\left\{\frac{1}{2}\right\}$; two roots, $\{-4 + 4\sqrt{3}, -4 - 4\sqrt{3}\}$

37. two roots, $\left\{\frac{3}{7}, 0\right\}$; one root, $\{-3\}$; no solution

38. 35 m wide, 40 m long   **41.** 8 in.

39. 9   **42.** 10 mph

40. about 4 sec   **43.** 25 sq cm

Pages 184–185

1. $(2, -1)$, downward; $(1, 3)$, upward; $(-3, -51)$, upward

2. $(2, -11)$, upward; $(2, 14)$, downward; $(-1, 7)$, downward

3. $(1, -1)$, downward; $(-1, -7)$, upward; $(4, -28)$, upward

4. vertex: $(2, -1)$ vertex: $(-2, -4)$

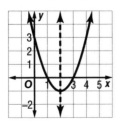
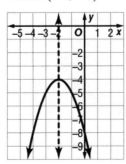

5. vertex: $(-2, 1)$ vertex: $(-1, 3)$

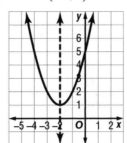
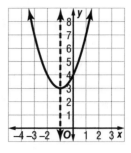

6. vertex: $(1, -3)$ vertex: $(1, -1)$

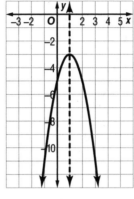
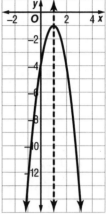

7. vertex: $(-2, 4)$ vertex: $(2, -1)$

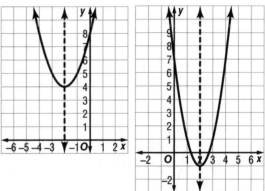

Page 189

1. G
2. B
3. H
4. J
5. A
6. D
7. E
8. I
9. C
10. F

1.

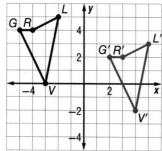

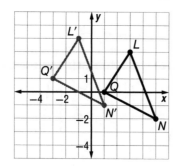

2.

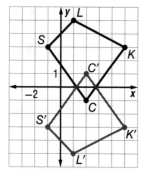

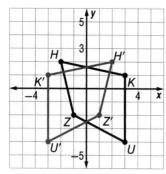

3.

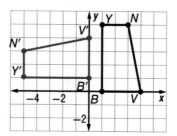

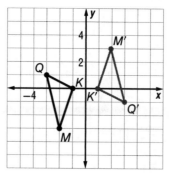

4. a

5. b

6. b

7. a

8. a

9. b

10. b

11. a

12. b

Pages 196–197

1. $(-14, 10)$, $r = 3$; $(-7, -6)$, $r = 6$; $(7, -11)$, $r = 5$

2. $(8, 5)$, $r = 2\sqrt{2}$; $(-6, -11)$, $r = 2\sqrt{13}$; $(-8, 15)$, $r = 2\sqrt{3}$

3. $\left(\frac{19}{2}, -\frac{9}{2}\right)$, $r = 4$; $\left(4\sqrt{7}, \frac{23}{2}\right)$, $r = 6$; $\left(\sqrt{35}, -\sqrt{13}\right)$, $r = 3\sqrt{6}$

4. center $= (-3, -1)$, radius $= 3$, $(x + 3)^2 + (y + 1)^2 = 9$; center $= (3, 4)$, radius $= 2$, $(x - 3)^2 + (y - 4)^2 = 4$

5. center $= (3, 0)$, radius $= 4$, $(x - 3)^2 + y^2 = 16$; center $= (-2, 4)$, radius $= 1$, $(x + 2)^2 + (y - 4)^2 = 1$

6. c

7. Area: 78.5 square units, Circumference: 31.4 units

8. b

Pages 198–203, Graphing Quadratics Review

1. $(2, 4)$, downward; $(-2, -3)$, upward; $(3, 3)$, upward

2. $(0, -5)$, upward; $(4, 3)$, downward; $(-4, 2)$, downward

3. $(1, -1)$, upward; $(-1, 2)$, downward; $(2, 5)$, downward

4. vertex: $(-2, 1)$ vertex: $(-1, -3)$

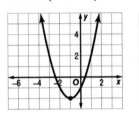 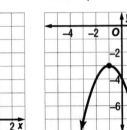

5. vertex: $(-1, -2)$ vertex: $(-2, -1)$

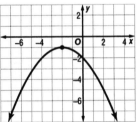

6. vertex: $(2, 0)$ vertex: $(3, -1)$

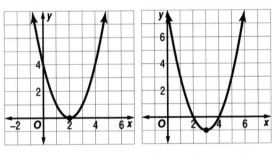

7. vertex: $(3, 4)$ vertex: $(-3, -2)$

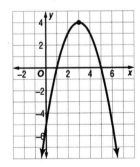

8. E
9. J
10. B
11. H
12. D
13. C
14. I
15. F
16. A
17. G

18.

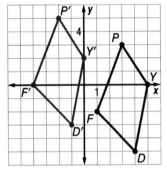

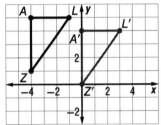

19.

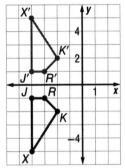

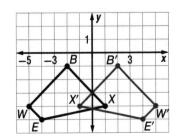

20.

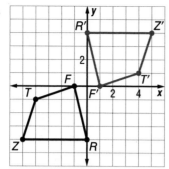

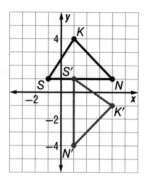

21. b

22. a

23. a

24. b

25. $(7, -2), r = 6;$
$(-11, -8), r = 7;$
$(-8, 1), r = 1$

26. $(1, -7), r = 3\sqrt{7};$
$(9, -16), r = \sqrt{3};$
$(-4, 8), r = 2\sqrt{6}$

27. $(0, 1), r = 8;$
$(7, 3), r = 3\sqrt{13};$
$(-3, 0), r = 2\sqrt{3}$

28. center $= (1, 1)$, radius $= 6$,
$(x - 1)^2 + (y - 1)^2 = 36;$
center $= (-4, -1)$, radius $= 3$,
$(x + 4)^2 + (y + 1)^2 = 9$

29. center $= (-1, -2)$, radius $= 4$,
$(x + 1)^2 + (y + 2)^2 = 16;$
center $= (3, 4)$, radius $= 1$,
$(x - 3)^2 + (y - 4)^2 = 1$

30.

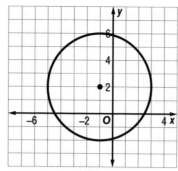

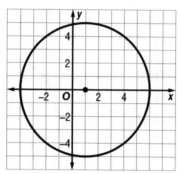

Pages 205–210

1. $-4, -20, -29$

2. $20, 4, 32$

3. $-54, -24, -6$

4. $1, -5, -14$

5. $\mathbb{R}, \mathbb{R}, \{x \mid x \neq -4\}$

6. $\{x \mid x \neq 2\}, \quad \{x \mid x \neq 5 \text{ and } x \neq -5\}, \quad \{x \mid x \leq 6\}$

7. $\{x \mid x \geq 4\}, \quad \{x \mid x \neq -3 \text{ and } x \neq 2\}, \quad \mathbb{R}$

8. No, Yes, No

9. Yes, Yes, No

10. domain: $\{x \mid -4 \leq x \leq 3\}$,
range: $\{y \mid -2 \leq y \leq 4\}$;
domain: $\mathbb{R}$, range: $\mathbb{R}$

11. domain: $\{x \mid -3 < x < 4\}$, range: $\{y \mid y = -2\}$;
domain: $\mathbb{R}$, range: $\{y \mid y \geq -4\}$

12. $x^2 + 5x - 4, \ 5x - 7, \ -5$

13. $x^3 + 3, \ 2x + 4, \ x^2 + 7$

14. $2x^2 + 8x, \ 3x^3 - x^2 - 3x + 1, \ 4x^2 + x - 3$

15. $\frac{x^2 + 1}{2}, x \neq 0; \quad x - 4, x \neq -4;$

$\frac{1}{x - 4}, x \neq -3 \text{ and } x \neq 4$

16. $3x - 15$, $x^2 - 10x + 25$, $9x$

17. $9x^2$, $3x^2$, $x - 10$

18. $4x^2 - 12x + 10$, $x^2 - 14x + 50$, $2x^2 - 1$

19. $2x - 17$, $x^2 - 6$, $2x - 10$

20. $x^4 + 2x^2 + 2$, $4x - 9$, $x - 14$

Pages 212–213

1.

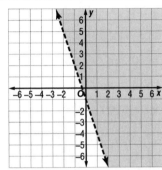

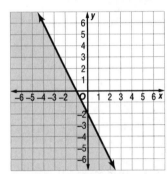

2.

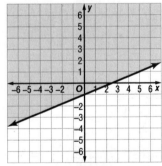

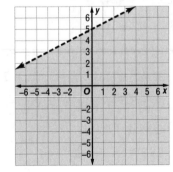

3.

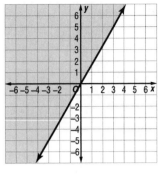

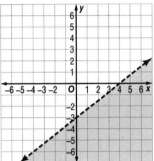

4.

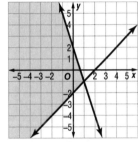

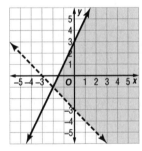

5.

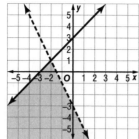

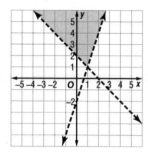

Pages 215–220

1. $\sin A = \frac{5}{13}, \cos A = \frac{12}{13}, \tan A = \frac{5}{12}$

   $\sin B = \frac{12}{13}, \cos B = \frac{5}{13}, \tan B = \frac{12}{5}$

2. $\sin E = \frac{4}{5}, \cos E = \frac{3}{5}, \tan E = \frac{4}{3}$

   $\sin F = \frac{3}{5}, \cos F = \frac{4}{5}, \tan F = \frac{3}{4}$

3. $\sin G = \frac{8}{17}, \cos G = \frac{15}{17}, \tan G = \frac{8}{15}$

   $\sin H = \frac{15}{17}, \cos H = \frac{8}{17}, \tan H = \frac{15}{8}$

4. $\sin K = \frac{\sqrt{2}}{2}, \cos K = \frac{\sqrt{2}}{2}, \tan K = 1$

   $\sin L = \frac{\sqrt{2}}{2}, \cos L = \frac{\sqrt{2}}{2}, \tan L = 1$

5. $\sin M = \frac{2\sqrt{10}}{11}, \cos M = \frac{9}{11}, \tan M = \frac{2\sqrt{10}}{9}$

   $\sin N = \frac{9}{11}, \cos N = \frac{2\sqrt{10}}{11}, \tan N = \frac{9\sqrt{10}}{20}$

6. 0.4226, 0.2588, 0.1584

7. 0.7071, 0.9848, 1.1106

8. 2.7475, 0.1219, 0.1736

9. 0.9455, 0.7771, 0.2679

10. 0.1564, 0.0698, 4.7046

11. 0.8090, 0.3839, 0.6691

12. 83°, 65°, 70°

13. 34°, 8°, 20°

14. 18°, 24°, 65°

15. 72°, 28°, 49°

16. 62°, 43°, 67°

17. 53°, 60°, 53°

18. 60°, 53°, 42°

19. 28°, 48°, 56°

20. $AC = 5$, $BC \approx 8.7$, $m\angle B = 30°$

    $DE \approx 3.6$, $EF \approx 2$, $m\angle D = 34°$

21. $AC \approx 15$, $AB \approx 18.3$, $m\angle B = 55°$

    $DF \approx 3.5$, $m\angle E = 60°$, $m\angle F = 30°$

22. $AB \approx 11.2$, $m\angle A \approx 63°$, $m\angle B \approx 27°$

    $DE \approx 3.2$, $DF \approx 4.4$, $m\angle D = 43°$

23. $AB \approx 15.3$, $m\angle A \approx 58°$, $m\angle B = 32°$

    $EF \approx 3.5$, $DF \approx 4.6$, $m\angle F = 41°$

Pages 221–226, Extension Topics Review

1. 34, −22, −4

2. 1, −14, 51

3. −5, 4, 11

4. −4, −4, 0

5. $\mathbb{R}$, $\mathbb{R}$, $\{x \mid x \neq 3\}$

6. $\{x \mid x \neq -6\}$, $\{x \mid x \neq 4 \text{ and } x \neq -4\}$, $\{x \mid x \leq 3\}$

7. $\{x \mid x \geq 3\}$, $\{x \mid x \neq -3 \text{ and } x \neq 1\}$, $\mathbb{R}$

8. No, Yes, No

9. No, Yes, Yes

10. domain: $\{x \mid -4 \leq x \leq 2\}$,
    range: $\{y \mid -4 \leq y \leq 2\}$;
    domain: $\mathbb{R}$, range: $\mathbb{R}$

11. domain: $\{x \mid -5 < x < 2\}$, range: $\{y \mid y = 3\}$;
    domain: $\mathbb{R}$, range: $\{y \mid y \leq 4\}$

12. $7x + 1$; $3x^2 - 8x - 3$; $-3x + 7$

13. $2x^2 + 3x - 4$; $x^2 - 2x + 1$; $\frac{1}{x}, x \neq 0 \text{ and } x \neq 2$

14. $6x^2 + 10x$; $\frac{1}{x+1}, x \neq -1 \text{ and } x \neq -5$; $5x - 2$

15. $x^3 - 5x^2 - 2x + 3$; $-3x^2 - 2x + 5$; $3, x \neq -3$

16. $2x^2 + 3$, $9x^2 - 12x + 3$, $6x + 1$

17. $4x^2 + 20x + 24$, $4x + 15$, $x^4 - 2x^2$

18. $6x^2 - 7$, $32x^2 + 80x + 50$, $12x + 8$

19.

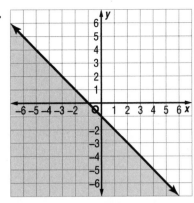

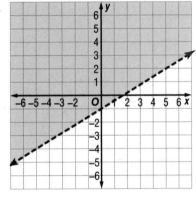

20.

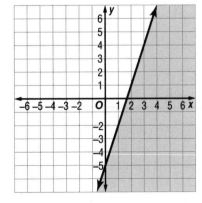

21.

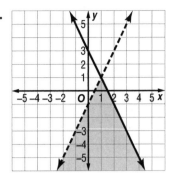

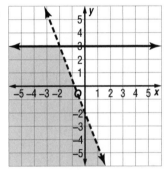

22.

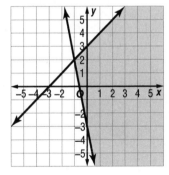

23. $\sin A = \frac{4}{5}$, $\cos A = \frac{3}{5}$, $\tan A = \frac{4}{3}$

$\sin B = \frac{3}{5}$, $\cos B = \frac{4}{5}$, $\tan B = \frac{3}{4}$

24. $\sin E = \frac{24}{25}$, $\cos E = \frac{7}{25}$, $\tan E = \frac{24}{7}$

$\sin F = \frac{7}{25}$, $\cos F = \frac{24}{25}$, $\tan F = \frac{7}{24}$

25. $\sin G = \frac{1}{2}$, $\cos G = \frac{\sqrt{3}}{2}$, $\tan G = \frac{\sqrt{3}}{3}$

$\sin H = \frac{\sqrt{3}}{2}$, $\cos H = \frac{1}{2}$, $\tan H = \sqrt{3}$

26. 46°, 10°, 16°

27. 25°, 68°, 66°

28. 84°, 39°, 50°

29. 37°, 32°, 63°

30. $AB \approx 20.7$, $AC \approx 16.1$, $m\angle A = 39°$

$DE \approx 13.0$, $DF \approx 11.9$, $m\angle D = 24°$

31. $AB = 17$, $BC = 12$, $m\angle A = 45°$

$DE \approx 20.6$, $m\angle D \approx 51°$, $m\angle E \approx 39°$

32. $AC \approx 7.3$, $m\angle A \approx 43°$, $m\angle B \approx 47°$

$DE \approx 2.5$, $DF \approx 1.5$, $m\angle D = 54°$

GLOSSARY

A

additive inverses A pair of opposite numbers whose sum is 0

additive property of equality A rule that you can add or subtract the same quantity on both sides of an equation. If $a = b$, then $a + c = b + c$.

area The measure of surface area of a figure

arithmetic average The sum of the values in a set divided by the number of values in the set

associative property A rule that states you can group numbers in any way when you are adding or when you are multiplying. Therefore, $(a + b) + c = a + (b + c)$.

B

base The number being multiplied by itself in an expression containing an exponent. In 2^5, 2 is the base.

binomial A polynomial with exactly two terms

C

canceling Crossing out the same factor in the numerator and the denominator of a fraction or a product of fractions without changing the value of the expression

circle The set of all points that are an equal distance, the radius, from a fixed point on the plane, the center

circumference The distance around a circle

commutative property A rule that states that you can add or multiply in any order. Therefore, $a + b = b + a$.

complementary angles Two angles whose measurements add up to 90°

completing the square A process for transforming a quadratic equation into the form of a perfect square equal to a positive number, and then solving the equation by taking the square root of both sides

composition of functions The process of combining two functions by substituting one function in place of each x in the other function. Therefore, $(f \circ g)(x) = f(g(x))$.

conic sections The family of curves including such shapes as the parabola and the circle

conjugates Two binomials whose only difference is the sign of one of the terms. The conjugate of $a + b$ is $a - b$.

consistent system A system of equations that has one solution

constant A number without a variable

cubed Raised to the 3rd power

D

dependent system A system of equations for which all real numbers are solutions

difference of squares A special factorization in which one perfect square is subtracted from another, as in $a^2 - b^2$

direct variation A relationship between two variables such that the ratio of the values of the variables always remains the same. If y varies directly as x, $y = kx$.

discontinuous function A function whose graph includes open circles, which are not part of the solution, and thus seems to jump from one spot to another on the grid

discriminant The expression in the radicand of the quadratic formula, $b^2 - 4ac$. For a quadratic equation of the form $ax^2 + bx + c$, the discriminant tells how many real roots exist.

distributive property A rule that states that for any numbers a, b, and c, $a(b + c) = ab + ac$

domain The set of real-number inputs in a function

E

elimination method A method for solving systems of equations by eliminating one variable by adding additive inverses

empty set A set with no members, shown { } or ∅

equation A mathematical statement that two expressions are equal

exponent A number that indicates how many times to use the base as a factor. The exponent is written as a superscript to the right of the base. In 2^5, 5 is the exponent and 2 is the base.

extraneous solution An answer that does not satisfy the original equation when solving a quadratic equation with two solutions

F

factoring Writing an expression as a product of its factors; applying the distributive property in reverse

formula An algebraic expression used to solve a problem

fractional exponent An exponent that is a fraction. In a relationship between a fractional exponent and a radical expression, the denominator of the fraction is the index number of the radical. The numerator is the exponent of the number inside the radical. $x^{\frac{2}{3}} = \sqrt[3]{x^2}$

function A relationship between a set of ordered pairs such that no two output values are produced by the same input value

H

hypotenuse The side opposite the right angle in a right triangle

I

identity An equation that is true for all real numbers

image A final shape formed by applying a transformation rule to a preimage

imaginary number The square root of a negative real number

inconsistent system A system of equations that has no solution

indirect variation See inverse variation.

inverse variation A relationship between two variables such that the product of the values of the variables always remains the same. If y varies inversely as x, $y = \frac{k}{x}$.

isolating the variable The process of using arithmetic operations to get the variable all alone on one side of the equation, with everything else on the other side

isosceles triangle A triangle with two equal sides and two equal angles

L

like terms Terms that have exactly the same variables raised to exactly the same powers

line of symmetry An imaginary line that divides a shape in half and is used to check for symmetry

M

mean See arithmetic average.

monomial A polynomial with exactly one term

multiplicative property of equality A rule that you can multiply or divide by the same nonzero number on both sides of an equation. If $a = b$, then $ac = bc$ and $\frac{a}{c} = \frac{b}{c}$.

N

non-rigid transformation A transformation rule that stretches or skews a preimage by changing the steepness of the curves

O

order of operations System for evaluating an expression. The order of operations specifies that you perform operations in parentheses first; then evaluate all exponents and roots; then perform multiplication and division from left to right; and finally, perform addition and subtraction from left to right.

P

parabola A unique shape that represents the graph of a quadratic equation

parallel lines Lines that lie in the same plane and never intersect

perfect square The product of an integer multiplied by itself

perimeter The distance around a figure

perpendicular lines Lines that intersect to form 90° angles

pi A constant value of approximately 3.14, used to find circumference and area of a circle

point-slope form The form of a linear equation written as $y - y_1 = m(x - x_1)$, where m = slope and (x_1, y_1) is a known point on the line

polynomial An algebraic expression that is a sum of terms

power An expression formed by a base and an exponent, or the value of such an expression

preimage An original, preliminary shape that forms a final shape when a transformation rule is applied

Pythagorean theorem A rule that states that in a right triangle, the sum of the squares of the lengths of the legs is equal to the square of the length of the hypotenuse. Therefore, for a triangle with sides a and b and hypotenuse c, $a^2 + b^2 = c^2$.

Q

quadratic formula A formula used to solve a quadratic equation of the form $ax^2 + bx + c$. The quadratic formula is $x = \frac{-b \pm \sqrt{b^2 - 4ac}}{2a}$.

R

radical expression An expression with at least one term that contains a radical

radicand The part of a radical expression that is written under the radical sign

range The set of real-number outputs in a function

rational expression A quotient that contains polynomials in both the numerator and the denominator

rationalizing the denominator Rewriting a fraction that contains radicals in its denominator to include only rational numbers in its denominator. To rationalize the denominator of a fraction, multiply the numerator and the denominator by the radical in the denominator. $\frac{\sqrt{2}}{\sqrt{5}} \times \frac{\sqrt{5}}{\sqrt{5}} = \frac{\sqrt{10}}{5}$

real number All values represented on a number line

reflection A transformation rule that flips a preimage across an axis, line, or point, creating a mirror image of the preimage

right angle An angle that measures 90°

rise The vertical change between two points on a line

root A solution to an equation

rotation A transformation rule that turns the preimage about a certain point

rules of equality Statements of what can and cannot be done to maintain balance in an equation

run The horizontal change between two points on a line

S

scientific notation Representation of a value as a product of two factors. The first factor is 1 or a number between 1 and 10, and the second factor is a power of 10 written in exponential form.

similar Figures in which all corresponding angles are equal and corresponding sides are proportional. The sign for similar is ~. $\triangle ABC \sim \triangle DEF$

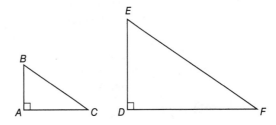

slope The measure of steepness or incline of a line

slope-intercept form The form of a linear equation solved for y, written as $y = mx + b$, where $m = $ slope and $b = y$-intercept

squared Raised to the 2nd power

standard form of a linear equation Form of a linear equation written with the variables in order on the left side of the equation and the constant on the right side. The coefficient of the x-term is a non-negative number.

standard form of a polynomial Form of a polynomial with terms written so that exponents descend, going from left to right

straight angle An angle that measures 180°

substitution method A method for solving a system of equations by substituting an expression that equals a variable in one equation in place of that variable in another equation in the system

supplementary angles Two angles whose measurements add up to 180°

symmetric A rule of equality that states that you can switch sides of an equation. If $a = b$, then $b = a$.

system of equations Two or more equations connected in such a way that the variables represent the same values in both equations

T

term A number or the product of a number and one or more variables raised to a power

transitive A rule of equality that states that you can substitute one equal value for another equal value. If $a = b$ and $b = c$, then $a = c$.

translation A transformation rule that applies changes to the x- and y-coordinates of a preimage's vertices such that the final image slides up, down, left, or right

trigonometry A branch of mathematics that studies the properties of trigonometric functions such as sine, cosine, and tangent

trinomial A polynomial with exactly three terms

U

uniform motion When the speed of an object stays constant over a period of time

V

variable A letter used to represent an unknown value

vertex The lowest or highest point on a parabola

vertical line test A test used to determine whether a graph represents a function. If any vertical line drawn on the graph crosses the graph at one point at most, then the graph represents a function. If a vertical line can be drawn such that it crosses the graph at more than one point, then the graph does not represent a function.

W

work problem A problem that examines how long it takes a number of workers to do a job and asks how long it will take if the number of workers increases or decreases. To solve a work problem, use the formula *work rate* × *time* = *work done*.

X

x-intercept The point at which a graph crosses the x-axis

Y

y-intercept The point at which a graph crosses the y-axis

Z

zero-product principle A rule stating that if the product of two factors is known to be 0, then one or both of the factors must be 0. Therefore, if $ab = 0$, then $a = 0$ or $b = 0$, or both.

INDEX